Jürgen Bock

Ontology Alignment using Biologically-inspired Optimisation Algorithms

Ontology Alignment using Biologically-inspired Optimisation Algorithms

Ontology Alignment using Biologically-inspired Optimisation Algorithms

by
Jürgen Bock

Dissertation, Karlsruher Institut für Technologie (KIT)
Fakultät für Wirtschaftswissenschaften, 2012
Tag der mündlichen Prüfung: 12. Juli 2012
Referenten: Prof. Dr. Rudi Studer, Prof. Dr. Hartmut Schmeck

Impressum

Karlsruher Institut für Technologie (KIT)
KIT Scientific Publishing
Straße am Forum 2
D-76131 Karlsruhe
www.ksp.kit.edu

KIT – Universität des Landes Baden-Württemberg und
nationales Forschungszentrum in der Helmholtz-Gemeinschaft

KIT Scientific Publishing 2013
Print on Demand

ISBN 978-3-86644-936-7

To my parents.

Preface

Looking outside our windows we recognise the amazing results of an enduring and massively parallel process called evolution. In billions of years evolution has created cells, plants, animals, and humans, including the human brain. With the ability to think, communicate, combine, and reason, the human brain is capable of dealing with complex information artifacts we call knowledge. While knowledge on the one hand is the root of all great inventions such as aeroplanes, pharmaceuticals, and computers, it also creates new challenges. The vast amounts of knowledge become difficult to manage, which is why we seek for assistance in computer science. One such difficulty in knowledge management is the fact that knowledge is distributed, which, in turn, leads to *ad hoc* or permanent knowledge integration tasks. In those cases where knowledge is represented formally in terms of ontologies, this problem has become known as *ontology alignment*.

Astonishingly, the emergence of parallel and distributed computing infrastructures now stimulates the imitation of evolutionary processes in the form of nature-inspired problem solving algorithms. The application of evolutionary and other biologically-inspired algorithms, such as computational swarm intelligence, have been shown very useful for complex optimisation problems. This book presents the work of my doctoral thesis on applying biologically-inspired optimisation algorithms to the problem of ontology alignment. It is a fascinating thought to take one of the first steps towards computationally assisting knowledge management by emulating the natural processes that created our brains in the first place.

This work would not have been possible without the support and encouragement of a number of people. My biggest thanks go to my doctoral adviser Prof. Dr Rudi Studer, who was providing an excellent research environment, allowing also somehow unconventional approaches to be explored. Also I'd like to thank Prof. Dr Hartmut Schmeck for acting as a second reviewer, and for giving valuable feedback. Many thanks also to Prof. Dr Stefan Nickel for being the examiner and to Prof. Dr Bruno Neibecker for chairing the examination.

For continuous motivation and expert advice throughout the process of developing this work I want thank Dr Stephan Grimm and Dr Sebastian Rudolph. Moreover, I'd like to thank Dr Catherina Burghart for holding (sometimes distracting) project work off me, particularly during the last months of writing up this document.

I wouldn't even have started this research without two people I met during my studies at Griffith University in Brisbane, Australia: Dr Andrew Lewis and Jan Hettenhausen. It was Andrew, who first brought the research area of biologically-inspired optimisation to my attention. With Jan I had a memorable brainstorming session about how to tackle the ontology alignment problem by Particle Swarm Optimisation, and endless nights of implementation, which eventually resulted in the first MapPSO prototype in 2008.

For fruitful discussions about biologically-inspired optimisation techniques and particularly about their application I'd like to express my gratitude to Dr Marcus Randall and Dr Sanaz Mostaghim.

For support with bringing the approach into "the cloud" my thanks go to Alex Lenk and Carsten Dänschel.

Many thanks go to Carsten Dänschel, Matthias Stumpp, Michael Mutter, Florian Berghoff, and Peng Liu for implementation support, fruitful discussions, and ideas for improving and optimising the prototypes. I also would like to thank the anonymous reviewers of submitted papers for their valuable feedback and comments.

Finally, I want to express my gratitude to my beloved parents Martin and Brigitte for their endless support and belief in me, and to Beáta Őri for her love and support.

Karlsruhe, July 2012
Jürgen Bock

Abstract

Ontologies describe real-world entities in terms of axioms, *i.e.* statements about them, and have become an established instrument for formally modelling and representing knowledge. The diversity of available ontologies results in a heterogeneous landscape where ontologies can overlap in their content. Such an overlap can be caused by ontologies modelling the same or similar domains created by different ontology designers, or with different views of a domain. If overlapping ontologies are to be used in a semantic application, sophisticated methods are required to overcome this heterogeneity. Identifying the overlap of ontologies is tackled by the discipline of *ontology alignment*.

An alignment between two ontologies denotes a set of correspondences between ontological entities. In this book, the ontology alignment problem is considered an optimisation problem. Thereby, optimality is defined in terms of an objective function that evaluates candidate alignments according to ontology modelling- and domain-specific criteria, *e.g.* significance and similarity of entity identifiers, or logical implications of an alignment. This optimisation problem is solved using biologically-inspired optimisation techniques, exemplary demonstrated by a novel Evolutionary Algorithm and an adapted Discrete Particle Swarm Optimisation algorithm. The Evolutionary Algorithm implements concepts from Evolutionary Programming and Extremal Optimisation and operates on a newly developed data structure for representing alignments. The Discrete Particle Swarm Optimisation algorithm extends an existing algorithm for a structurally similar problem.

The presented approach is the first to systematically apply biologically-inspired optimisation algorithms to the problem of ontology alignment. These algorithms have several advantages, which address relevant issues of the ontology alignment problem: First, the inherent parallelisability of biologically-inspired optimisation techniques enables the exploitation of distributed computing environments, such as cloud infrastructures. This improves on scalability aspects of the alignment task. Second, biologically-inspired optimisation algorithms are metaheuristics, which are largely independent from the objective function. Thus, arbitrary alignment quality criteria can be encoded, reflecting the characteristics of the ontologies. This makes the approach flexible regarding the nature of the ontologies. Third, candidate alignments are assessed as a whole during the optimisation process. This allows for consideration of global alignment quality criteria that go beyond the traditional pairwise computation of entity similarities. Finally, the iterative nature of biologically-inspired optimisation techniques demonstrates anytime behaviour, *i.e.* the algorithm can be interrupted at any time and the best alignment found so far can be obtained.

The presented algorithms were implemented in the form of two software prototypes, a generic ontology alignment API, and an evaluation library for flexibly building objective functions. The prototypes were evaluated using established ontology alignment benchmarks, among other experiments. It could be shown that biologically-inspired optimisation techniques are applicable to the ontology alignment problem and can compute alignments of good quality depending on the configuration of the objective function, while at the same time being scalable through high parallel efficiency.

Contents

List of Tables

List of Figures

List of Algorithms

1. Introduction

This work contributes to the state of the art by applying techniques from the discipline of Computational Intelligence to a problem in the area of Knowledge Representation. The problem under consideration is well-known under the name of *ontology alignment*. Here, ontology alignment is regarded an optimisation problem, and population-based *biologically-inspired optimisation techniques* are used for solving it.

The remainder of this chapter presents a motivation in Section 1.1 arguing that the presented approach is promising, and provides an overview of this book in Section 1.2.

1.1. Motivation

Today's world is characterised by an increasing amount of information being available to an increasing number of people, facilitated by technologies that provide or exploit global interconnectedness. For instance, the size of the Google™ search index increased from 1 billion (1,000,000,000) in the year 2000 to 1 trillion (1,000,000,000,000) Web pages in 2008[1]. In order to utilise and exploit these vast amounts of information efficiently, automatic systems are being developed to search, filter, combine, and interlink pieces of information to higher-level knowledge artifacts. In the research field of Knowledge Representation there have been enduring efforts in developing methods to symbolically represent complex statements and facts about arbitrary domains of interest in so-called *ontologies*. The Web Ontology Language (OWL) [92], for instance, provides means to represent knowledge about a domain of interest using a language with clearly defined logical underpinning. This provides real added value by allowing for inferring implicit knowledge from explicitly given axioms and facts, exploiting the formal semantics of the underlying logic. Semantic technologies are being developed in order to facilitate the creation

---

[1] `http://googleblog.blogspot.de/2008/07/we-knew-web-was-big.html` (accessed April 13, 2012)

and usage of ontologies, for instance, in semantic applications that benefit from that added value.

It is practically infeasible to have one large and global formal knowledge base that serves all purposes. Rather what can be observed is the emergence of a multitude of ontologies modelling specialised domain knowledge and serving specialised needs. However, there are use cases where incorporating knowledge from different existing ontologies is beneficial. Overcoming this heterogeneity in the ontology landscape by identifying overlaps between ontologies and creating correspondences between ontological objects is tackled in the discipline of *ontology alignment*.

Example. In the biomedical domain ontologies are widely used to represent complex correlations of anatomy, genetics, diseases, *etc.* Related pieces of information are used in different contexts, which resulted in the development of a multitude of ontologies. The BioPortal [96] is a coordinated effort to provide an infrastructure for hosting, maintaining, searching, and browsing these ontologies. In April 2012, the BioPortal hosted 297 ontologies, which are describing 6,365,010 terms, which are in turn used for 1,958,459,267 direct annotations[2].

In order to understand the basic concepts of ontology alignment, consider the two simple ontologies from the bibliographic domain shown in Figure 1.1. Both ontologies were developed by academic institutions in order to represent scientific publications. In this simplified visualisation the nodes represent classes of objects, and the tree structure expressed by indentation denotes subclass relationships (*is-a* relations). From the class labels it can be seen that both ontologies model various publication types, where Ontology 1.1b covers a wider domain than Ontology 1.1a, *i.e.* the ontologies have a partial overlap. The challenge for automatic ontology alignment algorithms is to identify this overlap and deliver an alignment, which is a set of correspondences, in this case between ontology classes. The correct alignment in this example seems obvious to an English speaking human being since he or she can make sense of the class labels. It might become more difficult if a language other than English is used, labels are expressed as synonyms, are of a technical jargon, or are omitted at all in favour of automatically generated identifiers. In the latter case, for instance, the labels cannot be used at all to compute an alignment, so other criteria such as hierarchy structures and other ontology features

---

[2]http://bioportal.bioontology.org/
(accessed April 13, 2012)

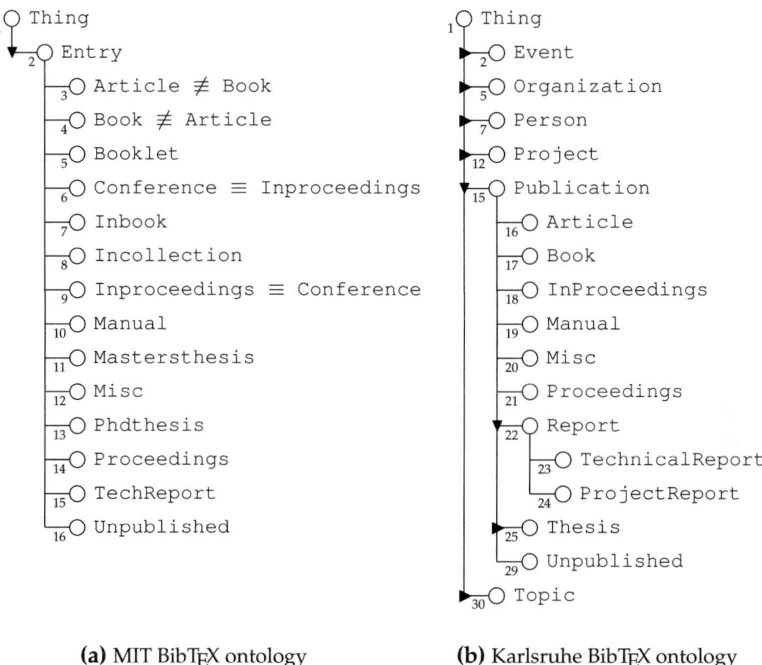

(a) MIT BibTEX ontology   **(b)** Karlsruhe BibTEX ontology

Figure 1.1.: Two example ontologies about the domain of bibliography. The figures show the class hierarchy of each ontology, which is only partially displayed/unfolded. Indented classes are subclasses of their parent, which denotes an *is-a* relationship between sub- and superclasses. Note that there is only a partial overlap between the two ontologies, since Ontology (b) covers a wider domain than Ontology (a).

need to be exploited. Abstracting from the concrete criteria that might be available in a particular case, there is a theoretically large number of possible alignments for any two ontologies. Given this large number of potential alignments the problem of ontology alignment can be seen as the problem of finding the best one among those candidates.

When it comes to complex optimisation problems, *nature* demonstrates astonishing solutions. By means of evolution, nature has produced creatures and organisms that are optimally adapted to their environments. Remarkable examples are insects that look like leafs for optimal camouflage, or carnivore plants that nourish on insects in order to survive in infertile environments. Evolutionary biology has revealed that nature finds optimal solutions by evolutionary processes, such as having a large population of individuals, reproduction, genetic recombination, mutation, and natural selection [35, 44].

Research in the area of Computational Intelligence is inspired by the way nature is solving problems. The discipline of Evolutionary Computation mimics evolutionary processes as found in natural systems, to solve complex optimisation problems. Several computational paradigms have been developed in this context, ranging from *Genetic Algorithms* to *Evolution Strategies*, which have been successfully applied to various real-world optimisation problems, *e.g.* in the engineering domain [36].

One particular result of natural evolution is the phenomenon of collaboration among population members, also known as *swarm intelligence*. Here, groups of individuals collaborate to solve problems that cannot be solved by singular individuals alone. Flocking birds or schooling fish, for instance, distract predators by appearing in large numbers, thus minimising the risk for each individual to be caught. Another example are ants communicating via pheromones in order to guide fellow individuals to food sources.

A more recent development in Computational Intelligence is the area of computational swarm intelligence, where the social behaviour of swarming animals is modelled. The most well-known paradigms are *Particle Swarm Optimisation* and *Ant Colony Optimisation*. Advocates of the swarm intelligence paradigm consider the incorporation of the social behaviour and the resulting information exchange between individuals as a real advantage over earlier techniques explored in the field [75].

In this book the term "biologically-inspired optimisation techniques" is used as an umbrella expression for all aforementioned natural phenomena transferred to computational models in order to solve complex optimisation problems. Also, the term "metaheuristics" is often used in order to express the universal applicability of the described techniques.

The interesting properties of biologically-inspired optimisation techniques and the importance and complexity of the ontology alignment problem motivated this research that strives for answering the following central question:

> *Can biologically-inspired, population-based optimisation algorithms be used to solve the ontology alignment problem?*

We expect that applying biologically-inspired optimisation techniques to the ontology alignment problem is a feasible and suitable approach. This expectation is based on the following conjectures regarding ontologies and ontology alignment, and the properties of biologically-inspired optimisation techniques that match those conjectures:

Conjecture 1 (*Scalability through Parallelisation*)**.** Due to increasing popularity and adoption in data intensive application domains, ontologies are continuously becoming larger in size. For instance, the Gene Ontology (GO) is updated daily and the number of represented "biological process terms" increased from 13,916 in September 2007 [125] to 22,382 in March 2012[3]—an increase of 60 % in 4.5 years.

Biologically-inspired optimisation algorithms are typically population-based. This feature makes them inherently parallelisable, since all computations for an individual in the population can be done independently from all other individuals. Typically, the costs for computing an individual are relatively high compared to the communication costs between individuals, which suggests high parallel efficiency according to Ahmdal's Law [1]. Hence it is expected that using biologically-inspired optimisation techniques for ontology alignment can demonstrate a gain in scalability.

Conjecture 2 (*Flexibility through Generic Metaheuristics*)**.** Despite the availability of ontology design guidelines and methodologies, *e.g.* ontology design patterns [101], there is only an imprecise notion of what constitutes a valid ontology. This results in a plethora of existing ontologies that reveal different characteristics in terms of how knowledge is modelled. While some ontologies, for instance, are modelled following a strict formal and axiomatic approach, others reflect a simple taxonomy with almost all concept meanings remaining implicit with the natural language semantics of concept labels. When computing an alignment between ontologies, those ontology characteristics play an important role when identifying corresponding ontology entities.

From the point of view taken in this book biologically-inspired optimisation techniques can be subsumed under the general term *metaheuristics*. Blum and Roli summarise that "metaheuristics are strategies that 'guide' the search process [...] and may make use of domain-specific knowledge in the form of heuristics that are controlled by the upper level

---

[3]`http://www.geneontology.org/GO.downloads.ontology.shtml` (accessed March 31, 2012)

strategy" [10]. For the application domain of ontology alignment, these heuristics can exploit any characteristics of the particular ontologies to be aligned. This makes biologically-inspired optimisation techniques flexible and largely independent of ontology characteristics. Once the chosen metaheuristic is tailored towards the ontology alignment problem, what remains is a matter of adjusting the heuristics according to the ontologies at hand.

Conjecture 3 (*Alignment-Level Optimisation through Global Evaluation*). The expressive power of state-of-the-art ontology modelling languages allows for the expression of complex statements and entity descriptions in ontologies. Particularly in the presence of logical negations, class descriptions in ontologies can become unsatisfiable, *i.e.* no instance can be assigned to the class without causing a logical contradiction. Ontologies containing unsatisfiable class descriptions are called *incoherent* [102]. In case correspondences as parts of an alignment are interpreted as statements about entity equivalence, such unsatisfiable class descriptions can be implied by the alignment. However, unsatisfiability is usually not caused by a single correspondence, but rather by a combination of two or more correspondences. For this reason, when selecting correspondences to be contained in an alignment, they cannot only be considered in isolation but globally in the context of the complete alignment. For example, consider the two ontologies shown in Figure 1.1. An alignment between them containing the correspondences a:`Article` $\leftrightarrow$ b:`Article` and a:`Book` $\leftrightarrow$ b:`Publication` would cause an incoherency, because a:`Article` and a:`Book` are disjoint in Ontology 1.1a and b:`Publication` and b:`Article` are in a subsumption relation in Ontology 1.1b. The incoherency cannot be detected by looking at each of the two correspondences in isolation.

Most traditional alignment approaches select correspondences according to the similarity of the corresponding entities based on the entities' local characteristics. Hence, such inter-correspondence effects causing incoherencies are usually ignored. Biologically-inspired optimisation techniques do not face this problem, since they evaluate solutions as a whole. Apart from assessing the contributions of solution components, they naturally consider their interplay.

In this respect there is a relation between the ontology alignment problem and the abstract NK model of rugged fitness landscapes introduced by Kauffman [73, 74]. The model describes the fitness landscape of a system of N interdependent (discrete) variables. The state of each variable contributes to the fitness of the entire system. The extent of the contribution depends on the state of the variable itself, as well as the state of other variables, it depends on. The number of dependencies is denoted as K. Applied to the problem of ontology alignment, the alignment represents the system, and the correspondences represent the variables. Kauffman explains the shape of the fitness landscape in terms of K, and concludes that a clear peak in the landscape can be observed for small K, while for large K, the fitness landscape demonstrates a rugged, almost random shape. Due to this observation it is difficult for standard greedy optimisation algorithms to determine the optimal configuration of the system, bearing the need for probabilistic, heuristic approaches, such as biologically-inspired optimisation techniques.

Conjecture 4 (*Anytime Behaviour through Iterative Approximation*). In several ontology processing tasks a "perfect" answer is not as important as getting an approximate answer quickly. This holds in particular if the loss in quality is compensated by the gain in runtime performance. This trade-off has been investigated for ontology reasoning [112], but not systematically analysed for ontology alignment. Depending on the application scenario, approximate solutions might be tolerable. For instance, in many cases of search result enhancements, where ontology alignments are exploited, a perfect alignment might not be as important as quick response time. On the other hand, in the case of aligning large ontologies in a periodic but not time-critical fashion (for instance in a nightly alignment computation) result quality prevails over short computation time.

In order to find optimal solutions, biologically-inspired optimisation algorithms perform an *iterative* search. In each iteration the best solution found so far can be obtained. This feature is called *anytime* behaviour and is intrinsic to biologically-inspired optimisation techniques. It depends on the use case to exploit this feature or wait until the algorithm terminates. The level of approximation is given by the elapsed portion of the maximum number of iterations to perform.

State-of-the-art approaches for solving the ontology alignment problem can only partially cope with the requirements described in these conjectures. Since most alignment algorithms involve the computation of similarity matrices between all pairs of ontology entities, they have problems taking into account global alignment constraints. The size of these similarity matrices grows quadratically with the size of the ontologies, thus hampering scalability. Furthermore, the matrix-based approach does not allow for delivering intermediate results in terms of anytime behaviour.

Paulheim [100] proposes a modularisation approach in order to improve scalability. To this end he splits the ontologies into overlapping fractions and computes partial alignments using standard methods. The modularisation, however, cannot consider global alignment constraints as described above.

1.2. Overview

This book is organised in eight chapters, including this introduction. The remainder is structured as follows.

Chapter 2 introduces the basic notions and foundations required in later chapters. In the first part of the chapter, Section 2.1 introduces ontologies and Section 2.2 introduces ontology alignment. In the second part of the chapter, biologically-inspired optimisation techniques are presented in Section 2.3, with a particular focus on population-based meta-heuristics. The most important definitions introduced in this chapter were published by the author at the *7th International Symposium on Foundations of Information and Knowledge Systems* (FoIKS 2012) [19].

Chapter 3 presents related work and prior art from various directions. Section 3.1 discusses related ontology alignment approaches in terms of their capabilities regarding global alignment criteria. To this end, an abstract categorising point of view is taken. Section 3.2 discusses concrete related approaches of applying biologically-inspired optimisation techniques to the problem of ontology alignment, in the area of semantic technologies in general, and to other problems that are structurally similar to the ontology alignment problem. Summarisations of the most relevant prior art were published by the author at the *7th International Symposium on Foundations of Information and Knowledge System* (FoIKS 2012) [19] and in the *Information Sciences* journal [16].

In Chapter 4, a number of evaluation metrics are presented that are used to compile an objective function for the optimisation algorithms. Thereby, Section 4.1 formally introduces similarity metrics on the correspondence level, as well as on the alignment level. In Section 4.2 several aggregation methods are presented in order to combine those correspondence or alignment evaluation scores. The presented metrics were continuously extended and improved throughout the development of the implementation prototypes and were partially, *i.e.* in earlier stages, published in the *Information Sciences* journal [16].

The core contribution of this work is presented in Chapter 5, namely the development of an Evolutionary Algorithm and a Discrete Particle Swarm Optimisation algorithm for the ontology alignment problem. To this end, an objective function used for assessing candidate alignments is defined in Section 5.1. Suitable encodings for representing alignments in the algorithms are introduced in Section 5.2. Section 5.3 describes the two algorithms and how they handle the iterative solution updates. Section 5.4 concludes the chapter with a discussion about several design choices regarding the algorithms and a comparison. The Evolutionary Algorithm presented in this chapter was published by the author at the *7th International Symposium on Foundations of Information and Knowledge System* (FoIKS 2012) [19]. The Discrete Particle Swarm Optimisation algorithm was published in the *Information Sciences* journal [16].

Implementation aspects are covered in Chapter 6. Apart from introducing the prototypes MapEVO in Section 6.3 and MapPSO in Section 6.4 a novel generic alignment API, named *KADMOS*, is introduced in Section 6.1. Since similarity metrics for correspondences and alignments are used in both prototypes in the same way, an independent software module called *HARMONIA* Commons is introduced in Section 6.2. The use of the *KADMOS* API and the *HARMONIA* Commons module is by no means restricted to the prototypes MapEVO and MapPSO. A focus in this chapter is also on the deployment of the prototypes, presented in Section 6.5, especially emphasising the issues related to deployment and exploiting parallelism in cloud computing infrastructures. All software prototypes are publicly available as open-source projects, and the MapPSO implementation prototype was partially described in the *Information Sciences* journal publication [16]. A dedicated paper describing the cloud deployment and the challenges faced with improving parallel efficiency was published by the author at the *5th International Workshop on Ontology Matching* (OM-2010) [17].

Chapter 7 describes evaluation results obtained by using the MapEVO and MapPSO prototypes. Since it is difficult to evaluate the various facets of the presented approach, several experiments have been conducted to cover the different aspects. Section 7.1 describes the performance metrics that are used for assessing alignment quality. Section 7.2 presents results from the continuous participation of MapPSO and MapEVO in the Ontology Alignment Evaluation Initiative (OAEI), as well as standalone experiments using the OAEI data sets. The diversity in result quality and its correlation to the instantiation of the objective function is studied in Section 7.3. Iterative convergence and the related feature of anytime behaviour is demonstrated in Section 7.4. An evaluation of the scalability by exploiting the parallelisability of population-based optimisation algorithms is presented in Section 7.5 by aligning two large biomedical ontologies on a cloud infrastructure. Section 7.6 concludes the chapter with a discussion about the evaluation results. The OAEI results were published by the campaign organisers in their summary papers at the *Ontology Matching* workshops from 2008 till 2011 [28, 45, 46, 47] as well as in the accompanying OAEI papers by the author [15, 18, 12, 13]. Some detailed OAEI results were further published in the *Information Sciences* journal [16]. Cloud-based scalability experiments were published at the *5th International Workshop on Ontology Matching* (OM-2010) [17] and at the *7th International Symposium on Foundations of Information and Knowledge System* (FoIKS 2012) [19]. The latter publication also contains experimental results using the OAEI data sets outside the official campaign modalities.

Chapter 8 summarises this work and its main findings in Section 8.1, and concludes by presenting an outlook on promising further research directions in Section 8.2.

2. Foundations

Applying a specific technique to a specific problem requires a basic understanding of relevant definitions, terms, and methods. This chapter presents fundamental notions required for understanding the main contributions of this work.

According to the twofold nature of this research, the chapter is divided into two parts. First, Sections 2.1 and 2.2 introduce foundations with respect to ontologies and ontology alignment. Thereby, the ontology alignment problem is considered as an optimisation problem, such that it can be targeted by biologically-inspired optimisation techniques, which are subject of Section 2.3. In this second part of the chapter, several population-based approaches are presented using a generic notation.

This chapter is not meant to serve as a comprehensive, textbook-like introduction to the research fields of ontology alignment or biologically-inspired optimisation. The interested reader is referred to the large body of introductory literature in the fields of ontology alignment [50], semantic technologies in general [65], or biologically-inspired optimisation techniques [3, 4, 75].

2.1. Ontologies

In the literature there is a wide range of definitions and unspoken agreements of what an ontology is. In 1993 Gruber defined an ontology as an "explicit specification of a conceptualization" [59]. This popular and frequently cited definition was later refined by Studer *et al.* [122] by the properties "formal" and "shared" in order to account for machine readability and generality aspects. In the context of ontology alignment the property of "explicitness" is of particular importance, since an algorithm identifying overlapping parts of ontologies relies on how explicit the specification of entities in the ontologies really is. A study [14] of the characteristics of ontologies found on the Web, for instance, revealed that a majority of the ontologies do not exploit expressive ontology language features to describe their entities. Instead, those ontologies model simple taxonomies with a significant portion of the semantics hidden in the natural language semantics of entity identifiers.

There are two frequently encountered points of view regarding the primary contents of an ontology. Firstly, an *axiom-centric* point of view, where an ontology is seen as a set of axioms that refers to a set of entities. Secondly, an *entity-centric* point of view, where an ontology is seen as a set of entities with additional axioms that make statements about the relationships among those entities. The key difference is that in the former point of view, an entity does not exist without any axiom referring to it, while in the latter, axioms are not necessarily required.

In the ontology alignment community typically the entity-centric point of view is common [50], however, in this book, ontologies are considered from the axiom-centric point of view, which is common in the logic community. The reason is the fact that axioms are the only way to express relations between and statements about entities, which is essential information for computing ontology alignments. An entity without any axioms referring to it would not carry any information apart from its identifier.

An ontology is represented in an ontology language. There have been many proposals for ontology languages in the past [31], however, few have evolved into dominating standards, such as RDF Schema [61], or the Web Ontology Language (OWL) [92]. The following definitions are abstract in terms of not being bound to any particular ontology language.

Definition 2.1. An *ontology* $\mathcal{O}$ is a set of axioms.

Ontology languages provide a range of language features. The notion of *expressiveness* applies to ontologies in terms of the expressiveness of the underlying ontology language, or in terms of the way language features are exploited by the ontology. Informally, an ontology is said to be of *low expressiveness* if its underlying ontology language is of low expressiveness, or if it exploits only a few language features. Conversely, an ontology is said to be of *high expressiveness* if its underlying ontology language is of high expressiveness, and it exploits a large number of those language features.

Definition 2.2. The *vocabulary* $\mathrm{voc}(\mathcal{O})$ of an ontology $\mathcal{O}$ is the set of entities, which are referred to by the axioms of $\mathcal{O}$. Every entity $e \in \mathrm{voc}(\mathcal{O})$ is associated with an *entity type* $\tau(e)$ with $\tau : \mathrm{voc}(\mathcal{O}) \to \mathcal{T}$, where $\mathcal{T}$ is a fixed finite set of types. For any given $t \in \mathcal{T}$ the *type-restricted vocabulary*

$\text{voc}_t(\mathcal{O}) = \{e \in \text{voc}(\mathcal{O}) \mid \tau(e) = t\}$ is the set of all entities in the vocabulary of type t.

Typical state-of-the-art ontology formalisms support vocabularies with various entity types of different semantics.

Example. In OWL the set of entity types is defined as

$$\mathcal{T} = \{\texttt{class}, \texttt{object\_property}, \texttt{data\_property}, \texttt{individual}\}$$

where, informally, an `individual` represents a real-world object, a `class` denotes a set of individuals, an `object_property` denotes a binary relation between two individuals, and a `data_property` denotes a binary relation between an individual and a data value.

Due to its foundation in Description Logics, OWL 2 DL [99] features three types of axioms: TBox axioms (terminology), ABox axioms (assertions) [2], and RBox axioms (roles) [68].

Definition 2.3. Let $\mathcal{O}$ be an ontology. An *annotation property* in $\mathcal{O}$ is a function $a : \text{voc}(\mathcal{O}) \rightarrow 2^{\texttt{AnnotationValue}}$, mapping entities to sets of annotation values, where an annotation value can be "a literal [*data value*, author's remark], an IRI, or an anonymous individual" [92]. The *annotation set* $\text{annot}(\mathcal{O})$ of an ontology $\mathcal{O}$ is the set of all functions a, where a is an annotation property occurring in $\mathcal{O}$.

The fact that for an entity there can be multiple annotations using the same annotation property requires the function to deliver a set of annotation values. In contrast to entities in the vocabulary of an ontology, annotation properties have no semantics and are not considered as entities in this book. Consequently annotation properties cannot participate in correspondences and alignments. However, since annotations frequently encode (non-logic) natural language descriptions of entities depending on the usage scenario of the ontology, they provide important information in order to determine similarities between entities.

Example. In OWL a frequently used annotation is `rdfs:label` providing a label describing an entity, typically in natural language.

Definition 2.4. The *size* of an ontology $\sharp\mathcal{O}$ is the number of entities referred to by its axioms. Hence, $\sharp\mathcal{O} = |\text{voc}(\mathcal{O})|$.

For every $t \in \mathcal{T}$ let $\sharp_t\mathcal{O} = |\text{voc}_t(\mathcal{O})|$ denote the number of ontology entities of a specific type t, also called the *size of $\mathcal{O}$ with respect to t*.

2.2. Ontology Alignment

The discipline of *ontology alignment* tackles the problem of heterogeneity in the ontology landscape. In this book, the term "ontology alignment" (or "alignment" for short) is used for both the process of identifying overlapping parts of ontologies, and the result of this process in terms of a collection of correspondences. Thereby, an alignment spans *two* ontologies and considers ontology *entities* as participating objects.

The following sections formalise ontology alignment as required in the subsequent chapters, and introduce the alignment problem as an optimisation problem.

2.2.1. Alignment Formalism

The notion of ontology alignment requires the notion of a correspondence, which is defined as follows.

Definition 2.5. Given two ontologies $\mathcal{O}_1$ and $\mathcal{O}_2$, a *correspondence* between $\mathcal{O}_1$ and $\mathcal{O}_2$ is a pair of entities $C = \langle e, f \rangle$, where $e \in \mathrm{voc}_t(\mathcal{O}_1)$ and $f \in \mathrm{voc}_t(\mathcal{O}_2)$, *i.e.* e and f are of the same entity type $t \in \mathcal{T}$. The set of all possible correspondences is $\mathcal{C} = \bigcup_{t \in \mathcal{T}} \mathrm{voc}_t(\mathcal{O}_1) \times \mathrm{voc}_t(\mathcal{O}_2)$, the set of all pairs of entities in $\mathcal{O}_1$ and $\mathcal{O}_2$ with matching types.

The definition includes the restriction that correspondences need to be type specific, *i.e.* a correspondence cannot hold between two entities of different types. This definition determines the constraints for valid correspondences, but does not allow for comparing any two correspondences.

Definition 2.6. The *confidence* of a correspondence is defined as a function $\iota : \mathcal{C} \to [0, 1]$, denoting a certainty of a correspondence. To this end, a confidence of 0 means least certainty, while a confidence of 1 means highest certainty.

Arbitrary quality metrics can be used to calculate the confidence of a correspondence, allowing for evaluation, comparison, or selection.

The correspondence definition allows for the definition of an alignment as follows.

Definition 2.7. Given two ontologies $\mathcal{O}_1$ and $\mathcal{O}_2$, an *alignment* $A \subseteq \mathcal{C}$ between $\mathcal{O}_1$ and $\mathcal{O}_2$ is a set of correspondences between entities of $\mathcal{O}_1$ and $\mathcal{O}_2$. An alignment is called *valid* is it satisfies the following conditions:

- For each $e \in \text{voc}(\mathcal{O}_1)$ there is at most one $f \in \text{voc}(\mathcal{O}_2)$ such that $\langle e, f \rangle \in A$.

- For each $f \in \text{voc}(\mathcal{O}_2)$ there is at most one $e \in \text{voc}(\mathcal{O}_1)$ such that $\langle e, f \rangle \in A$.

A valid alignment thus constitutes an injective, functional relation. The set of all possible alignments is $\mathcal{A} = \{A \mid A \in 2^{\mathcal{C}}, A \text{ is a valid alignment}\}$, where $2^{\mathcal{C}}$ denotes the power set of all possible correspondences.

For the sake of brevity in the remainder of this book, when referring to an alignment it is considered a valid alignment, unless stated otherwise.

Definition 2.8. The *size* of an alignment A is the number of correspondences it contains, denoted as $|A|$.

Note that the conditions in Definition 2.7 restrict the size of an alignment to be less or equal the size of the smaller ontology:

$$|A| \quad \leq \quad \sum_{t \in \mathcal{T}} \min\{\sharp_t \mathcal{O}_1, \sharp_t \mathcal{O}_2\} \quad \leq \quad \min\{\sharp \mathcal{O}_1, \sharp \mathcal{O}_2\} \qquad (2.1)$$

Definition 2.9. The alignment *quality* is defined as a function $F : \mathcal{A} \rightarrow [0, 1]$, assigning an evaluation score to an alignment. To this end, a quality of 0 means the lowest (worst) evaluation score, while a quality of 1 means the highest (best) evaluation score.

Typically, the correspondence confidence ι and the alignment quality F are related in a way that the quality of an alignment computes from the confidence values of its correspondences plus additional alignment evaluation metrics. However, this relation is not enforced by either definition.

The literature provides several generalisations of these definitions that allow for more expressive correspondences and alignments. Euzenat and Shvaiko [50] summarise in details the developments in the field. One typical generalisation on the alignment level is to relax (or abolish) the restrictions in Definition 2.7. This can allow for an entity to participate in more than one correspondence, or force every entity of an ontology to participate in at least one correspondence. This characteristic is denoted as *alignment multiplicity* (*alignment cardinality*). Definition 2.7 used in this book allows for partial alignments, where there might be no complete overlap of ontologies, and enforces alignments to be precise, *i.e.* disallowing an entity of an ontology to correspond to more than one entity of the other ontology. (This is what Euzenat and Shvaiko call a "?:?" alignment [50].)

An extension on the correspondence level is the specification of a *relation* expressed by a correspondence. This relation between corresponding entities can be equivalence, specialisation/generalisation, disjointness, or general relatedness. In principle, all relations between entities possible in the given ontology language can be used as correspondence relation. In this book there is no distinction between different correspondence relations. Definition 2.7 deliberately does not define any semantics for correspondences and alignments. Indeed the interpretation of correspondences and alignments is strongly use case dependent. However, in many cases a correspondence between ontological entities is seen as a statement expressing that those entities are "equivalent" or at least in some sense "similar". Depending on the interpretation of correspondences, they can have an impact on the semantics of entities defined in the ontologies to the level of introducing incoherency or inconsistency. Irrespective of the correspondence interpretation in the use case at hand, a common assumption is to regard a correspondence as equivalence axiom for the two corresponding entities.

The following definitions use OWL axioms [92] in order to express ontology axioms induced by an alignment. (Note that this does not limit the argument to this particular ontology language, as axioms from other languages can be used analogously.)

Definition 2.10. Let $C = \langle e, f \rangle \in A$ be a correspondence in A. The *ontology axiom induced by C* is defined as

$$a_C = \begin{cases} EquivClasses(e, f) & \text{if } \tau(e) = \tau(f) = \texttt{class} \\ EquivObjectProps(e, f) & \text{if } \tau(e) = \tau(f) = \texttt{object\_property} \\ EquivDataProps(e, f) & \text{if } \tau(e) = \tau(f) = \texttt{data\_property} \\ SameIndividuals(e, f) & \text{if } \tau(e) = \tau(f) = \texttt{individual} \end{cases}$$

According to Definition 2.1 with an ontology being a set of axioms, a merged ontology can simply be created as the set union of the axioms from both ontologies.

Definition 2.11. Let $\mathcal{O}_1$ and $\mathcal{O}_2$ be ontologies, and let A be an alignment between them. The *merged ontology based on A* is defined as

$$\mathcal{O}_A = \mathcal{O}_1 \cup \mathcal{O}_2 \cup \{a_C \mid C \in A\}$$

Informally, the merged ontology based on an alignment A is the union of the set of axioms from the two ontologies aligned by A, augmented by the ontology axioms induced by the correspondences in A. Note that $\mathcal{O}_A$ is itself an ontology.

The notions from Definitions 2.10 and 2.11 are relevant since the effects of adding axioms induced by an alignment can be used to define the correspondence confidence and alignment quality functions.

2.2.2. Ontology Alignment Problem

The *ontology alignment problem* is the problem of finding an optimal alignment for two given ontologies. An alignment A between two ontologies $\mathcal{O}_1$ and $\mathcal{O}_2$ is called *optimal* with respect to F iff there is no other alignment A' between $\mathcal{O}_1$ and $\mathcal{O}_2$, such that $F(A') > F(A)$. In other words, the optimal alignment A^* is defined as

$$A^* = \text{argmax}_{A \in \mathcal{A}}\, F(A) \tag{2.2}$$

which is the alignment of best quality according to F.

The *solution space* for this optimisation problem is the set of all possible alignments $\mathcal{A}$ as from Definition 2.7. With no prior assumptions about the input ontologies and the expected alignment, every alignment $A \in \mathcal{A}$, which is valid by definition, is a candidate solution to the ontology alignment problem.

Size of the Solution Space. Finding the optimal alignment is a challenging endeavour due to the size of the solution space. In order to determine the size of the solution space for two given ontologies $\mathcal{O}_1$ and $\mathcal{O}_2$, let $m = \sharp_t \mathcal{O}_1$ and $n = \sharp_t \mathcal{O}_2$ be the number of entities of the same type t in two ontologies $\mathcal{O}_1$ and $\mathcal{O}_2$. By Definition 2.7, a valid alignment can contain at most $\min\{m, n\}$ correspondences (for entities of type t). Consider the number of alignments with exactly $k \leq \min\{m, n\}$ correspondences. To fix one such alignment, one has to pick k elements from $\text{voc}_t(\mathcal{O}_1)$, and, independently, k elements from $\text{voc}_t(\mathcal{O}_2)$. For each of these choices there are $k!$ different ways of arranging the selected sets into k non-overlapping pairs from $\text{voc}_t(\mathcal{O}_1) \times \text{voc}_t(\mathcal{O}_2)$. Thus there are

$$\binom{m}{k} \cdot \binom{n}{k} \cdot k! = \frac{m!}{(m-k)!}\binom{n}{k} \tag{2.3}$$

alignments of size k. Consequently, the number of all alignments of arbitrary size is

$$\sum_{k=0}^{\min\{m,n\}} \frac{m!}{(m-k)!}\binom{n}{k} \tag{2.4}$$

To determine the total number of alignments, one has to calculate the number of possible combinations by multiplying the respective single-type alignment counts. Thus the number of total alignments is

$$\prod_{t\in\mathcal{T}} \sum_{k=0}^{\min\{\sharp_t\mathcal{O}_1,\sharp_t\mathcal{O}_2\}} \frac{\sharp_t\mathcal{O}_1!}{(\sharp_t\mathcal{O}_1-k)!}\binom{\sharp_t\mathcal{O}_2}{k} \tag{2.5}$$

The size of the solution space is exponential with respect to the size of the ontologies, since neither the size of the alignment, *i.e.* the degree to which the ontologies overlap, nor the assignment of entity pairs is known upfront. Figure 2.1 illustrates the growth of the solution space for a single entity type with growing ontology sizes.

This exponential size of the solution space implies that an enumeration of solutions or an exhaustive search for the optimal solution is infeasible for real-world ontologies. The study underlines the challenge imposed by ontologies of increasing sizes as stated in the introductory Conjecture 1.

2.3. Biologically-inspired Optimisation Methods

Located in the research area of *Computational Intelligence*, biologically-inspired methods are algorithms that mimic natural phenomena by an artificial computational model. Particularly interesting phenomena that can be observed are optimisation processes, such as the optimal adaptation of organisms to their natural environment, or the collaborative behavioural patterns of hunters or prey.

Following the model of nature, most biologically-inspired optimisation techniques are population-based, randomised algorithms [75]. This requires a notion for random numbers as follows.

Definition 2.12. $\text{rand}_U \in [0, 1]$ defines a real-valued uniform random number between 0 and 1.

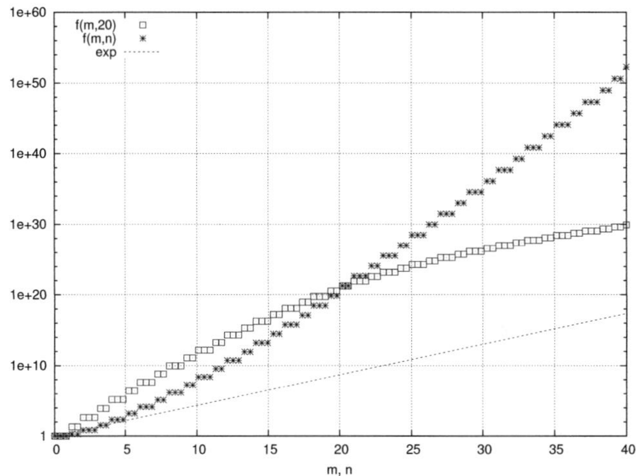

Figure 2.1.: Size of the solution space (logarithmic scale), considering only a single entity type with $f(m, n)$ representing Equation (2.4). With $m = \sharp_t \mathcal{O}_1$ and $n = \sharp_t \mathcal{O}_2$, $f(m, 20)$ shows the growth of the solution space with constant $n = 20$, while $f(m, n)$ illustrates the growth for $m = n$ equally sized ontologies. ($\exp = e^m$ as comparison).

The function $\mathrm{rand}_U : 2^{\mathbb{N}} \rightarrow \mathbb{N}$ selects an element from a set of natural numbers with uniform probability. That is, given a set $\{1, \ldots, n\}$, $\mathrm{rand}_U(\{1, \ldots, n\})$ selects a natural number between 1 and n with probability $1/n$.

For the sake of simplicity, rand_U denotes both a real-valued number and a function. In the remainder it will always be clear from the context in which manner it is used.

This section presents the classical representatives [44, 75, 43] of nature-inspired optimisation techniques in their basic versions[1]. Without doubt there are many variations of these algorithms designed for specific problems, in particular regarding the various probabilities and update operations involved (cf. Chapter 3 for related examples). Moreover, different

---

[1]The reader familiar with biologically-inspired optimisation literature might find the notations used here non-standard. However, the notation used in this book can cover all mentioned optimisation paradigms in an abstract manner and allows for specialising the algorithms for the ontology alignment problem by minimal adjustments.

algorithms are traditionally designed for either continuous or discrete optimisation problems. However, there are variants of all algorithms making them applicable to both continuous and discrete problems.

The focus of this book is on population-based optimisation metaheuristics. As a single example for a biologically-inspired optimisation technique that is not population-based, Extremal Optimisation is mentioned.

Disregarding the particularities of any tangible algorithm, population-based optimisation metaheuristics can be generally introduced as follows.

Definition 2.13. Let $\mathbb{P}$ be the problem space associated to some problem. A *population* is defined as a pair $\langle I, p \rangle$, where I is a finite set of *individuals* and $p : I \to \mathbb{P}$ is a function assigning to every individual of the population a position in the problem space.

Definition 2.14. An *update operation* is a function U mapping populations to populations.

Definition 2.15. Let U be an update operation. An *optimisation run* with respect to U is a finite sequence $\langle I_1, p_1 \rangle, \ldots, \langle I_n, p_n \rangle$ of populations, such that $\langle I_{i+1}, p_{i+1} \rangle = U(\langle I_i, p_i \rangle)$ for all $i \in \{1, \ldots, n-1\}$. The numbers $1, \ldots, n$ are referred to as *iterations*[2].

The following paragraphs use these abstract definitions and describe the differences of various metaheuristics that have been proposed in the past. Most notably these differences refer to the update operation U that controls the repositioning of individuals and thus the convergence of the algorithm throughout the iterations.

2.3.1. Evolutionary Computation

Inspired by the Darwinian theory of the evolution of species [35], *"evolutionary computation* refers to computer-based problem solving systems that use computational models of evolutionary processes, such as natural selection, survival of the fittest and reproduction, as the fundamental components of such computational systems." [44, Chapter 8]. Optimisation algorithms that employ Evolutionary Computation techniques are commonly classified as *Evolutionary Algorithms* [44, Chapter 8].

---

[2]In Evolutionary Algorithms, typically the term "generation" is preferred over "iteration" due to its direct analogy in natural systems.

The following paragraphs describe the basic variants of Evolutionary Algorithms in the light of the general notations of population-based optimisation metaheuristics given in the beginning of this section.

Genetic Algorithms

Probably the most popular and widely known class of Evolutionary Algorithms are *Genetic Algorithms*, first introduced by Holland [66]. The paradigm simulates sexual reproduction with solutions represented as chromosomes (individuals). Typically, a binary encoding is used for potentially continuous parameter ranges, thus resulting in a separation between the genotype (individual representation) and the phenotype (solution representation).

Given a population $\langle I, p \rangle$, an individual $x \in I$ (chromosome) represents a binary encoding of a parameter configuration as solution candidate. In each iteration (generation) during the optimisation run, the update operation U creates a new population of different individuals, with the population size remaining constant. This is done in several steps:

1. Given the parent population $\langle I_i, p_i \rangle$, for every individual $x \in I_i$ a fitness value is computed, reflecting the quality of the solution represented by the individual with respect to the objective function.

2. From the parent population $\langle I_i, p_i \rangle$, $|I_i|$ independent selections are performed (*e.g.* using "roulette wheel" selection) in order to obtain a temporary population $\langle I_{i'}, p_{i'} \rangle$ of individuals that are allowed to reproduce. The selection is based on the fitness scores, such that better adapted individuals have a higher probability to be selected (possibly multiple times).

3. From the selected individuals random pairs are chosen for reproduction. A given probability determines whether this reproduction step will involve the crossover operation. There are various variations of the crossover operator (n-point, uniform, *etc.*), sharing the general idea of exchanging genetic material between individuals. Formally, the temporary population $\langle I_{i'}, p_{i'} \rangle$ from the previous step is transformed into another temporary population $\langle I_{i''}, p_{i''} \rangle$ as follows: Based on a given reproduction rate, a subset of $I_{i''}$ is created by crossover operations from randomly selected parents from $I_{i'}$. Retaining the original population size, remaining individuals for

$I_{i''}$ are selected randomly from the parent population $I_{i'}$, such that $|I_{i''}| = |I_i|$.

4. From the new temporary population $\langle I_{i''}, p_{i''} \rangle$ individuals undergo a random mutation with a given, though generally small probability. Mutation in this case means flipping bits on the bit string of an individual with the given probability. After this operation, the population in the new iteration is $\langle I_{i+1}, p_{i+1} \rangle$ with $|I_{i+1}| = |I_i|$.

Evolution Strategies

Developed by Rechenberg [108] and Schwefel [115], *Evolution Strategies* is a paradigm for optimising parameters of an objective function with a focus on self-adaptation of its evolution process. Thus, in analogy with nature one can speak of "evolution of evolution" [44, 75]. In contrast to Genetic Algorithms, parameters are encoded directly as real-valued numbers, which makes Evolution Strategies best suitable for continuous optimisation problems.

Given a population $\langle I, p \rangle$, an individual $x \in I$ represents a parameter configuration as solution candidate, as well as evolution parameters that influence the update operation. In each iteration during the optimisation run, the update operation U creates a new population of different individuals, with the population size remaining constant. In order to obtain the new generation, for a population $\langle I_i, p_i \rangle$, U creates a temporary population $\langle I_{i'}, p_{i'} \rangle$, with $\mu = |I_i|$ being the number of *parents* and $\lambda = |I_{i'}|$ being the number of *offsprings*. Offsprings are created by applying recombination and mutation, utilising the evolution parameters being part of each individual's representation. There are two types of selection strategies in order to determine the new population $\langle I_{i+1}, p_{i+1} \rangle$:

- The (μ, λ) selection selects the best μ individuals from the λ offsprings only.

- The $(\mu + \lambda)$ selection selects the best μ individuals from the union of the μ parents and the λ offsprings.

In order to increase selection pressure, the number of offsprings to be generated temporarily is recommended to be about seven times the size of the parent population [43, Chapter 4].

Evolutionary Programming

First introduced by Fogel [52], *Evolutionary Programming* simulates evolution by maintaining a population of individuals that are exposed to the environment (objective function), and favouring those that are best adapted (survival of the fittest).

As opposed to Genetic Algorithms, the update function in Evolutionary Programming does not involve a recombination operation for exchanging information between individuals in the population. However, each individual creates an offspring by mutation, temporarily doubling the size of the population.

Formally, the update operation U for Evolutionary Programming is defined as follows: In a temporary population $\langle I_{i'}, p_{i'} \rangle$, for each $j \in \{1, \ldots, |I_i|\}$ let $x_{|I_i|+j} \in I_{i'}$ be the mutated species created by x_j. Thus, for all $k \in \{1, \ldots, |I_{i'}|\}$

$$p_{i'}(x_k) = \begin{cases} p_i(x_k) & \text{if } 1 \leq k \leq |I_i|, \\ q_i(x_k) & \text{if } (|I_i| + 1) \leq k \leq |I_{i'}| \end{cases}$$

where q_i maps a species to the new position after mutation.

In a subsequent selection step, half of that population becomes extinct, returning to the original size of the population. Survivors are typically determined by some sort of tournament selection, where species pairwise compete with each other, which results in a ranking that is used to select survivors of that iteration. Formally, using a particular selection principle, the population in the $(i + 1)$th iteration is $\langle I_{i+1}, p_{i+1} \rangle$, such that $I_{i+1} \subseteq I_{i'}$ with $|I_{i+1}| = |I_i|$.

Differential Evolution

A relatively new population-based Evolutionary Computation paradigm is *Differential Evolution* developed by Storn and Price [121]. Traditionally, the algorithm is tailored to optimise real-valued objective functions, *i.e.* given a population $\langle I, p \rangle$, for every $x \in I$, $p(x) = (x_1, x_2, \ldots, x_n)$ is a vector of real-valued parameters. As update operation, the approach employs a mixture of mutation and crossover. Given a population $\langle I_i, p_i \rangle$ in iteration i, the update operator U generates the population of the $(i+1)$th iteration as follows: For every individual $x^t \in I$ with $t \in \{1, \ldots, |I_i|\}$, a *trial* individual is created, by selecting three individuals $x^{(1)}, x^{(2)}, x^{(3)} \in I_i$, such that $x^{(1)} \neq x^{(2)} \neq x^{(3)} \neq x^t$. Let $r = \text{rand}_U(\{1, \ldots, n\})$. The trial individual x' is composed as $p(x') = (x'_1, x'_2, \ldots, x'_n)$, such that for all $j \in \{1, \ldots, n\}$

$$x'_j = \begin{cases} x_j^{(3)} + \gamma(x_j^{(1)} + x_j^{(2)}) & \text{if } \phi_j < p_r \text{ or } r = j \\ x_j^t & \text{otherwise} \end{cases} \tag{2.6}$$

where $p_r \in [0, 1]$ is the probability of reproduction, $\phi_j = \text{rand}_U$, a new uniform random number for every j, and γ is a real-valued factor to control the impact of the differential variation. The second condition $r = j$ ensures that at least one parameter in x' is modified via recombination.

For the $(i + 1)$th iteration every individual $x^t \in I$ is replaced with the trial individual x' created following Equation (2.6) iff x' improves over x^t. Otherwise x^t is transferred into the next iteration unchanged.

There exists also a binary version for Differential Evolution, developed by Pampará *et al.* [98].

Extremal Optimisation

It can be observed that evolution in nature according to the principles of natural selection does *not* happen by systematic breeding of well-adapted species, but rather by extinction of poorly adapted ones. Motivated by this phenomenon, Boettcher and Percus [23, 24] developed a technique called Extremal Optimisation. Thereby, a solution to the given optimisation problem is developed iteratively by modifying solution components according to the Bak-Sneppen model [5] that describes evolution happening in avalanches even though the whole system experiences only small constant changes. Accordingly, in each iteration, only the worst performing solution component and its two neighbours are removed and replaced by random new components. Due to the fact that the solution is constantly changing, the algorithm never converges [106].

The classical version of Extremal Optimisation maintains a single solution that is iteratively modified as described above. Randall proposes an enhanced, population-based variant [105], which applies the same principles of the Bak-Sneppen model to a population of individuals, each representing a solution. To this end, at regular intervals throughout the optimisation run, the worst performing individual and its two closest neighbours in terms of shared solution components are removed and replaced by random new ones.

Genetic Programming

An evolutionary algorithm for a predefined application domain is *Genetic Programming* [79]. This application domain is the generation of a computer programme best suitable to solve a given problem. Thinking about Genetic Programming as a means to create an ontology alignment algorithm does not appear completely unrealistic, however, the approach would significantly differ from the one followed in this book. While this work investigates ways to discover an alignment between two ontologies that is optimal with respect to certain quality criteria, a Genetic Programming approach would strive for finding an algorithm that produces optimal alignments. Thus, Genetic Programming would tackle the problem at a higher abstraction level, which makes it irrelevant in the context of this work.

2.3.2. Computational Swarm Intelligence

In contrast to the area of Evolutionary Computation, *computational swarm intelligence* is inspired by the natural phenomenon of swarming animals, such as schooling fish or flocking birds. The most distinct feature in computational swarm intelligence is social behaviour in the sense that individuals in the population are influenced by other individuals when moving through the problem space [75].

The following paragraphs describe the two basic variants of computational swarm intelligence, Particle Swarm Optimisation and Ant Colony Optimisation, in the light of the general notation of population-based metaheuristics from the beginning of this section.

Particle Swarm Optimisation

Initially developed by Kennedy and Eberhart [76] and later refined by Shi and Eberhart [116] in the late 1990s, *Particle Swarm Optimisation* is a relatively young population-based optimisation paradigm. In contrast to Evolutionary Algorithms presented earlier in this section, the population of individuals (swarm) remains constant throughout the complete optimisation run. That is, no individual (particle) joins or leaves the swarm, no offsprings are created, and no individual becomes extinct. Instead, in each iteration[3] every individual moves to a new position in the problem space, influenced by other individuals in the population. Thus, for each iteration $i \in \{1, \ldots, n - 1\}$, $I_{i+1} = I_i$. In its classical implementation, every individual remembers the best position in the problem space it has ever visited (*personal best*) and knows about the best position any individual in its neighbourhood has ever visited (*neighbourhood best*). In each iteration i the update operation U adds a velocity vector $v_i(x_j)$ to the position $p_i(x_j)$ of each individual $x_j \in I_i$, $1 \leq j \leq |I_i|$, changing its position in the problem space. The velocity vector is composed of the personal and neighbourhood best positions, as well as an inertia component [116]. The new population in the $(i + 1)$th iteration thus is $\langle I_{i+1}, p_{i+1} \rangle = U(\langle I_i, p_i \rangle) = \langle I_i, p_{i+1} \rangle$, where $\forall x_j \in I_i, 1 \leq j \leq |I_i| : p_{i+1}(x_j) = p_i(x_j) + v_i(x_j)$.

There are different *neighbourhood topologies* that influence the social interaction between individuals, and thus how information about good positions in the problem space propagates through the population [44]. A straightforward variant is *gBest* Particle Swarm Optimisation that implements a complete graph topology, *i.e.* every individual knows about the personal best of every other individual, thus the neighbourhood best is always the best position the whole population has ever seen. On the other hand, *lBest* Particle Swarm Optimisation implements a neighbourhood topology, where an individual only knows about the personal best positions of some of the other individuals (its neighbours).

Ant Colony Optimisation

The optimisation algorithm originally introduced by Dorigo in his PhD thesis [39], called *Ant System* was the first of a family of algorithms that became known as *Ant Colony Optimisation* [41]. Inspired by the behaviour of natural ants, the optimisation problem needs to be encoded as a graph, which can be traversed by the artificial ants in order to find an optimal path. In this approach the social component, which is typical for the swarm intelligence paradigm, is realised by means of pheromone trails that influence the probability of ants choosing edges on the graph.

The most notable differences to Particle Swarm Optimisation is that the population members have no memory, and the social component is not

---

[3]In the field of Particle Swarm Optimisation, typically a more adapted terminology is used. So the "population" is called "swarm", "individuals" are called "particles", and "generations" are called "iterations".

realised by direct communication but indirectly by pheromone trails deposited on the graph—a principle called "stigmergy". Thus there is also no notion of a social network structure that governs the communication between groups of individuals (or subswarms).

Even though there is no need to keep the population size constant, due to the lack of individual memory or social networks, Ant Colony Optimisation algorithms maintain a constant number of individuals (ants). Thus, for each iteration $i \in \{1, \ldots, n-1\}$, $I_{i+1} = I_i$. Let the problem space be represented as a graph $G = (V, E)$, where V is a set of nodes and E is a set of edges. Let $s \in V$ be the start node, and let $t \in V$ be the target node. After traversing the graph from s to t in iteration $i \in \{1, \ldots, n\}$, every individual $x_j \in I_i, 1 \leq j \leq |I_i|$ represents a solution $p_i(x_j) = (s, v_1, \ldots, v_{n_j}, t)$, with $s, t, v_1, \ldots, v_{n_j} \in V$. Subsequently, every solution is evaluated and pheromones are updated for the edges. Depending on the concrete Ant Colony Optimisation variant, different updating strategies can be followed. Also depending on the Ant Colony Optimisation variant is the behaviour of an ant when traversing the graph in subsequent iterations. In this respect, the approaches differ in the way ants subsequently choose edges to follow. At each node an ant makes a probabilistic decision based on pheromone values and (greedy) local support heuristics.

Ant Colony Optimisation algorithms have been successfully applied to many graph problems, such as the Travelling Salesman Problem [40].

3. Related Work

The presented work brings together two research areas within the wide field of artificial intelligence: a problem in the area of *Knowledge Representation*, namely ontology alignment, is tackled by techniques from the area of *Computational Intelligence*, more precisely by means of biologically-inspired optimisation techniques. Due to this "intersubdisciplinary" nature, there are numerous research endeavours that can be regarded related to this one.

In the following Section 3.1 ontology alignment approaches are categorised and discussed from a rather generic point of view. It shall be noted that there is no complete overview of the ontology alignment literature presented, and the interested reader is referred to pertinent surveys that are available. However, ontology alignment approaches are categorised coarsely according to the way they tackle the ontology alignment problem. Particularly constraint-based approaches are highlighted in Section 3.1.2, since they can consider global alignment constraints and thus have an advantage over traditional matrix-based approaches, which are generically summarised in Section 3.1.1. The constraint-based approaches thus share the same advantage as the approach presented in this work.

Related work regarding the application of biologically-inspired optimisation techniques is presented in Section 3.2. In particular three main areas of related work are important to be discussed:

- Previous approaches of applying biologically-inspired optimisation techniques in the context of ontology alignment. These most relevant efforts are discussed in Section 3.2.1.

- Previous approaches of applying biologically-inspired optimisation techniques in the context of semantic technologies in general. This widens the scope of the problem domain and underlines the growing interest and relevance in the semantic technologies community. Prior work in this direction is presented in Section 3.2.2.

- Previous approaches of applying biologically-inspired optimisation techniques in the context of other problem domains that are structurally similar to the problem of ontology alignment. Earlier developed techniques and applications that are relevant and were inspiring for this work are presented in Section 3.2.3.

3.1. Ontology Alignment

Ontology alignment is an active field of research within the area of semantic technologies. Not only the growing interest in the Linked Data Web [9] establishes new challenging application areas for ontology alignment, but also the adoption and development of expressive ontologies in specialised domains such as bioinformatics [117, 21]. This increasing interest in solutions for the ontology alignment problem motivated the development of a large number of approaches. A detailed overview is given by frequent surveys [57, 50, 30], the pertinent Web resources of the ontology alignment research community[1], or via the Ontology Alignment Evaluation Initiative (OAEI)[2] [49].

The discipline of ontology alignment is closely related to the more general task of schema matching [8] or graph matching. Indeed, several techniques for entity similarity computation, such as similarity flooding [90] or virtual document construction [103] (cf. Chapter 4) have been developed for graph matching and transferred to ontology alignment.

Despite the large number of ontology alignment approaches there are only a few general paradigms of how the alignment problem is tackled. The most prominent and widely used paradigm is a *matrix-based* approach, where an alignment is extracted from a matrix reflecting pairwise entity comparisons. A second approach is *constraint-based*, where an alignment is generated in a way that it meets global validity and quality constraints. The following paragraphs provide more details about those two approaches.

---

[1]`http://ontologymatching.org/`
(accessed March 23, 2012)
[2]`http://oaei.ontologymatching.org/`
(accessed March 23, 2012)

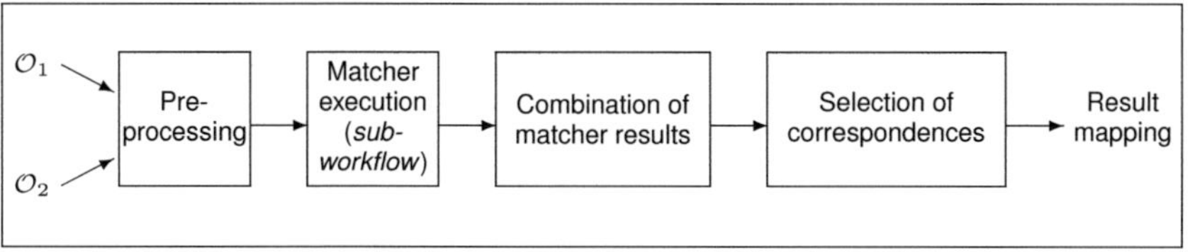

Figure 3.1.: General workflow for pairwise ontology matching (according to Rahm [104]).

3.1.1. Matrix-based Approaches

Most state-of-the-art ontology alignment systems follow a pairwise ontology matching approach by computing one or several similarity matrices. Hereby, a similarity matrix denotes a data structure containing similarity scores for all pairs of ontology entities from the two ontologies to be aligned. Figure 3.1 illustrates the typical workflow for pairwise ontology matching as described by Rahm [104]. The step "matcher execution" refers to the computation of one or more similarity measures for the entity pairs, which are combined before a set of correspondences (*i.e.* an alignment) is selected.

The numerous alignment approaches following this matrix-based principle differ in the details at the various steps along this generic workflow. Different similarity metrics, for instance, can be incorporated and used to compute a number of similarity matrices. Also, the combination of similarity metrics (matcher results) can be done in different ways. Finally, the extraction of alignments from the combined similarity matrices, which is typically a variant of the Hungarian method for assignment problems [80, 93] can vary across different alignment approaches, as for instance in the AgreementMaker system [34].

Matrix-based approaches typically have problems taking global alignment evaluation metrics into account. While a similarity can be derived for any entity pair considering the entities' local characteristics and possibly their ontology context, it is difficult to assess a correspondence in the presence of other correspondences. The reason is that the presence of any correspondence in the alignment is not known before the selection step (cf. Figure 3.1), but in order to incorporate it into the evaluation of a potential entity pair the information is required already when computing the matrix entries in the matcher execution step. Some state-of-the-art alignment systems came up with methods to deal with this problem.

The RiMOM (Risk Minimization based Ontology Mapping) system[3] uses Bayesian decision theory in order to select correspondences from the similarity matrices [124, 123]. This decision making process is based on the entities' characteristics and their ontology context, but disregards the alignment context (presence of other correspondences). While not being explicitly mentioned by the authors, this problem, however, is partially addressed by an iterative process allowing the extracted alignment to be refined and corrective actions to be taken.

The ASMOV system[4] [69] follows the same approach of an iterative alignment correction. In contrast to RiMOM the goal of this iterative process is explicitly dedicated to the semantic verification, *i.e.* the correction of alignments whose correspondences imply subsumptions in the merged ontology that cannot be verified by the ontologies themselves.

3.1.2. Constraint-based Approaches

In contrast to *matrix-based* approaches the *constraint-based* approaches consider the alignment problem as a constraint satisfaction problem. The most notable difference to matrix-based approaches is the ability to consider global alignment-level constraints that naturally account for the fact that correspondences might influence each other in terms of contributing to a good or bad alignment, and thus cannot be considered in isolation.

Ontology Merging using Answer Set Programming

An approach for merging ontologies by tackling the task as a constraint satisfaction problem was first introduced by the author in 2006 [11, 20]. An algorithm was presented using the *Answer Set Programming* formalism [54] to declaratively denote the constraints a valid merging has to fulfil. Apart from generic structural constraints the proposed algorithm utilises the capabilities of the answer set solver *dlvhex*[5] to evaluate the truth value of logic programming atoms externally. In the particular case

---

[3]`http://keg.cs.tsinghua.edu.cn/project/RiMOM/`
(accessed October 11, 2011)

[4]`http://www.infotechsoft.com/products/asmov.aspx`
(accessed March 28, 2012)

[5]`http://www.kr.tuwien.ac.at/research/systems/dlvhex/`
(accessed October 14, 2011)

of ontology merging, a *dlvhex* plugin was developed to accesses the lexical WordNet® database in order to exploit linguistic knowledge intrinsic to the labels in the ontologies. Consequently the answer set solver can compute answer sets, each one representing a valid merged ontology according to the given constraints.

Depending on the input ontologies several answer sets may be produced. On the one hand this provides the possibility to incorporate a human user in the way that he or she could simply select one of the "proposals" provided by the algorithm. On the other hand this shows that simple constraints, such as linguistic features based on WordNet® are often not sufficient to come up with a unique intuitive solution. However, due to the declarative nature of the constraint-based approach the answer set program could easily be augmented by more declarations that cover other characteristics of "intuitively" good solutions.

A drawback of the solution is that one could observe poor runtime performance in spite of the highly optimised implementation of the answer set solver used. This is due to the inherent combinatorial explosion when computing models in disjunctive logic programming formalisms, such as Answer Set Programming.

CODI

A more mature implementation of constraint-based ontology alignment is developed at the University of Mannheim, named CODI [95]. This system uses Markov logic to declaratively encode hard and soft constraints in order to restrict the alignment produced by the algorithm as follows:

- *hard constraints* are enforced during the computation of an alignment, *i.e.* there cannot be an alignment violating any of those constraints. Alignment cardinality, alignment coherency [88], or any other condition that needs to hold for a valid alignment can be enforced by hard constraints.

- *soft constraints* are not enforced but have weights attached, which represent the "importance" of the constraint. Structural ontology properties as well as so-called *a priori* similarities resulting from lexical entity comparison are examples for soft constraints provided by the authors.

After encoding all constraints as logic formulae, the problem of ontology alignment is solved by computing the most probable alignment using maximum *a posteriori* inferencing in the Markov logic framework. The authors use integer linear programming in order to efficiently perform this sort of inference problem [109].

3.2. Applications of Biologically-inspired Optimisation Methods

While biologically-inspired optimisation metaheuristics are making their way into applications in the engineering domain [36, 83], they have only recently raised interest in the field of semantic technologies. The topic is particularly pushed by a research group at the Vrije Universiteit Amsterdam, a recently organised workshop "NatuReS"[6] (Nature-inspired Reasoning for the Semantic Web), or the interest in special journal issues, such as the IEEE Computational Intelligence Magazine with a special issue "Semantic Web meets Computational Intelligence"[7].

This section surveys the existing approaches of applying biologically-inspired metaheuristics to the problem of ontology alignment and other areas of semantic technologies research. Additionally, the application of biologically-inspired metaheuristics in problem domains that are structurally similar to the ontology alignment problem are investigated.

3.2.1. Applications in Ontology Alignment

To date, biologically-inspired metaheuristics have rarely been applied in the context of ontology alignment. To the best of the author's knowledge, there is only one proposal, GAOM, that applies a Genetic Algorithm to the alignment problem by treating it directly as an optimisation problem as it is done by the approach presented in this book. Other applications of biologically-inspired metaheuristics, such as the systems GOAL and ECOMatch, treat the problem of parameter configuration for matching systems. Theses approaches are presented in the following paragraphs.

GAOM.   Proposed in 2006 by Wang *et al.* [130] the GAOM system applies a Genetic Algorithm to the ontology alignment problem. To this end, an

---

[6] `http://natures.few.vu.nl/2008/`
(accessed March 28, 2012)
[7] IEEE Computational Intelligence Magazine Volume 7, Issue 2

individual in the population (chromosome) represents a valid solution to the alignment between two ontologies $\mathcal{O}_1$ and $\mathcal{O}_2$. Let $n_1 = \sharp\mathcal{O}_1$ and $n_2 = \sharp\mathcal{O}_2$. A solution is represented as a string $m(1)\ m(2)\ \cdots\ m(n_1)$, where m is a function $m : \{1, \ldots, n_1\} \to \{1, \ldots, n_2\}$, such that $m(i)$ denotes a correspondence between the ith concept of $\mathcal{O}_1$ and the $m(i)$th concept of $\mathcal{O}_2$. This representation limits the algorithm to compute only alignments where every concept from $\mathcal{O}_1$ occurs in exactly one correspondence, while each concept in $\mathcal{O}_2$ can correspond to multiple concepts in $\mathcal{O}_1$. This lax definition of the solution string avoids the need for corrective crossover operations as no invalid solutions can be created. Moreover, the algorithm is only feasible when a complete alignment for one ontology is desirable, *i.e.* no partial overlap between the ontologies is expected.

The fitness function is determined by the number of elements in the solution string that actually match. Given a solution string two rules are applied to evaluate each correspondence. The first rule, called "intentional rule" checks for a lexical match of several local feature values of the concepts participating in the correspondence. These local features are concept names, "properties related with the concept", and "set of instances associated with the concept" [130]. However, the authors leave it unclear what exactly is meant by those names, properties, and instances, in particular with respect to the lexical check that is performed. The second rule, called "extensional rule" incorporates dependencies between different correspondences in a solution, thus introducing a global alignment evaluation metric that goes beyond the assessment of single correspondences in isolation. For two concepts $c_1, c_2 \in \mathcal{O}_1$ that correspond to concepts $c_1', c_2' \in \mathcal{O}_2$, respectively, the rule checks for a lexical match between the "relationships" between the two concepts within each ontology. Again, the authors leave it unclear what exactly is meant by those relationships.

Although the authors report positive evaluation results with respect to the OAEI 2005 *benchmarks* data sets, the algorithm did non officially participate in the campaign or at least provides detailed results other than aggregated evaluation scores. Unfortunately, the system is not publicly available and only presented in a single publication [130]. It seems it has not been maintained after that publication in 2006, since the authors are not responsive to requests and have been publishing in different fields of research since this publication. This leaves the conclusion that the status of GAOM is as presented in this single publication meaning that it is capable of aligning only *concepts* of two ontologies, where one ontol-

ogy is completely aligned with another one. This is sufficient for the
OAEI benchmark data sets, but inappropriate for real-world ontologies
that have only a partial overlap.

Alignment Parameter Optimisation. Contributing to the problem of
ontology alignment on the meta level, an application of Genetic Algorithms named "Genetics for Ontology Alignment" (GOAL) was introduced by Martinez-Gil *et al.* [86, 87]. In contrast to the GAOM system
the ontology alignment problem is not considered an optimisation problem *per se*, but as a problem of computing a weighted average of similarity
measures used to obtain the actual alignment. The authors focus on finding the optimal weight configuration for several similarity measures. To
this end, a Genetic Algorithm is applied in four independent optimisation
tasks. The fitness function used in these tasks coincide with alignment
precision, recall, F-measure, and number of false positives, respectively.

Indispensable to this optimisation is the presence of a reference alignment for the given ontology pair. This makes the approach at best useful
for two ontologies with no given reference alignment, if those ontologies have similar characteristics to other ontologies for which a reference
alignment is known. Depending on the characteristics the optimal weight
configuration could be transferred to the new pair of ontologies with un-
known alignment. Experiments reported by the authors show that for
the OAEI 2007 *benchmarks* data sets, very good results with respect to all
objective functions could be obtained.

Similar to the GOAL approach, Ritze and Paulheim present the ECO-
Match system [110] that tackles the problem of parameter optimisation at
a broader level. While GOAL only considers weights for similarity measures, ECOMatch optimises the complete set of parameters for any given
ontology alignment system. The alignment system itself is used as a black
box. The authors conduct experiments with various biologically-inspired
metaheuristics, *e.g.* Genetic Algorithms, Differential Evolution, and Hill
Climbing in order to optimise the F-measure with respect to a partial reference alignment. This partial reference alignment is a fraction of the
complete reference alignment provided by a domain expert. The authors
show that for a partial reference that covers at least 10 % of the complete
reference, the F-measure with respect to this partial reference correlates
with the F-measure with respect to the complete reference. Consequently,
such a partial reference is sufficient to determine a parameter configuration that is expected to be optimal even if no complete reference is avail-

able. These insights by Ritze and Paulheim regarding the sufficiency of a partial reference could be applied to the GOAL approach by Martinez-Gil *et al.* [86, 87] when optimising weights of similarity measures. However, no empirical evaluation in this direction has been done so far.

Another approach similar to GOAL, but with an exclusive focus on instance features, was proposed by Wang *et al.* [131]. This work aims at identifying concept correspondences based on the instances asserted to those concepts, where assertions do not necessarily need to be shared across the two ontologies. Among other techniques, the authors evaluate the feasibility of Evolution Strategies in order to identify the optimal weighting of instance features.

3.2.2. Applications in Other Semantic Technologies

Apart from their application in ontology alignment, population-based optimisation techniques have been utilised in other fields of semantic technologies with increasing popularity.

Oren *et al.* [97] use a Genetic Algorithm in order to tackle the problem of RDF query answering. With an RDF query being a graph pattern, the authors consider the single triple patterns and partial triple patterns as constraints to be satisfied by every correct query result. Consequently the fitness function is defined by the number of constraints being violated by a given solution candidate. Every individual in the population represents a valid solution, *i.e.* a complete variable assignment with respect to the graph pattern. Following the principles of Genetic Algorithms, individuals undergo selection, recombination, and mutation processes. The authors promote the approximate and anytime characteristics of their approach, which reflect the nature and requirements of the Semantic Web. In a subsequent work [60] the authors improve and evaluate their algorithm regarding performance and scalability. They report positive and encouraging results, also regarding the general feasibility of an approximate approach for the query answering problem.

The approach taken in this book is to consider a particular problem as optimisation problem and utilise biologically-inspired optimisation algorithms for solving it. Unlike this optimisation-based point of view, other approaches utilise "Swarm Intelligence" in a more general analogy, highlighting the social and collaborative aspects of independent and distributed agents.

Dentler *et al.*, for instance, tackle the problem of computing the deductive closure of a (possibly distributed) RDF graph [38]. In this case, distributed agents independently traverse the RDF graph and apply rules according to the underlying RDF Schema. Thus, implicit statements are materialised and made explicit by the agents. Different types of agents apply different rules, thus keeping the single individuals simple. Despite the analogy to ants it would be misleading to talk about an Ant Colony Optimisation algorithm, since the approach does not search for an optimal path or solve any other optimisation problem as done by typical Ant Colony Optimisation systems.

One should note that the term "Swarm Intelligence" is sometimes used in a rather sloppy manner, for instance to denote a collection of different agents pursuing different jobs in an adaptive information retrieval architecture on the Web. Ratnayake *et al.* [107], for instance, describe how a small set of four agents acts as a modularised software architecture. This is not reflecting the general understanding in Computational Intelligence and in this book, accordingly. Hence such approaches are not considered relevant for this work.

3.2.3. Applications in Structurally Related Problem Domains

Abstracting from the concrete problem of ontology alignment, this section surveys applications of biologically-inspired metaheuristics to problems which are structurally similar to ontology alignment. In particular these are approaches for *discrete* optimisation problems, as the alignment problem can be seen as the problem to either select or not select any potential correspondence for being part of the solution, *i.e.* the optimal alignment.

Discrete Particle Swarm Optimisation for Attribute Selection

Although traditionally designed and best suitable for continuous optimisation problems, Particle Swarm Optimisation (PSO) has also successfully been applied to discrete problems.

Correa *et al.* [32, 33] introduced a novel discrete version of the Particle Swarm Optimisation algorithm, (DPSO) and applied it to the problem of attribute selection for a Naïve Bayes classifier. The objective is to find the smallest set of attributes, which maximises the predictive accuracy of the classifier. Unlike classical binary PSO [77] the proposed DPSO does not represent solutions as bit strings of equal length, but rather as sets of indexes, where set sizes can vary from particle to particle. While the position in the search space is represented as this set of attribute indexes, the velocities are represented as "proportional likelihoods". The particle update is done in three steps:

1. Determine proportional likelihoods for attribute indexes. Let n be the number of possible attributes, and let $|I|$ be the number of particles in the swarm. Let $x_i \in I$ be a particle, with $i \in \{1, \ldots, |I|\}$, and $p(x_i) \subseteq \{1, \ldots, n\}$ the attribute selection represented by x_i. Further, let $p_{\text{personal}}(x_i)$ be its personal best, and p_{global} be the global best. For each attribute index $j \in \{1, \ldots, n\}$ compute a proportional likelihood as follows:

   a) Assign an initial proportional likelihood of 1.

   b) Add a constant α, iff j is contained in $p(x_i)$.

   c) Add a constant β, iff j is contained in $p_{\text{personal}}(x_i)$.

   d) Add a constant γ, iff j is contained in p_{global}.

   e) Multiply with a uniform random number $\varphi_j \in (0, 1)$.

2. Sort attribute indexes according to their proportional likelihoods.

3. Select the k attribute indexes with the highest proportional likelihood, where k is the size of particle x_i.

When comparing the DPSO with the classical binary PSO on a classification task using a bioinformatics data set, the authors report similar scores with respect to predictive accuracy. However, regarding the number of attributes selected, the DPSO identifies a smaller set, which is preferable. The reason for this is the initialisation of particles. In the binary PSO, each bit in the bit string is initially set to 1 with a probability of 0.5, which results in a distribution[8] of particles sizes, which is centred around $n/2$, where n is the number of available attributes. Apart from the criterion of minimising the number of attributes, both approaches (binary PSO and DPSO) increase the prediction accuracy compared to the use of all available attributes.

---

[8]For n attributes, there are n Bernoulli trials with a success (selection) probability of $p = 0.5$, resulting in a binomial distribution.

Assignment Type Problems

Related to the problem of ontology alignment are all types of assignment problems. Randall *et al.* [106] investigate the application of Extremal Optimisation for the *General Assignment Problem*, the *Bin Packing Problem*, and the *Capacitated Single Allocation Hub Location Problem*. While the canonical Extremal Optimisation algorithm could not provide satisfactory results, an enhanced population-based approach [105] can deal with constraints that determine the feasibility of solutions found. To this end, the approach allows the transition through infeasible areas of the solution space. Moreover, the incorporation of *local search* is reported as a crucial component in order to improve the solution after the coarse-grained search of the metaheuristic.

Sequence Alignment in Molecular Biology

A problem that is also structurally similar to ontology alignment is the problem of aligning DNA, RNA, or amino acid sequences in the context of molecular biology. As in ontology alignment there is typically no complete overlap between sequences due to insertions, replacements, or deletions occurring in the nucleotide or amino acid sequences. Chellapilla and Fogel [29] have used Evolutionary Programming for solving the multi-sequence alignment based on their observation that dynamic programming approaches traditionally used for this problem reveal bad performance when the number of sequences is large, the average lengths of the sequences is large, and similarities between the sequences are low. The Evolutionary Programming algorithm the authors propose makes use of several mutation operators that are adapted to the problem at hand. To this end, "shuffle" operators perform swapping of symbols or subsequences, and "growing" and "recombination" operators compute variations that take into account the presence of already aligned columns.

3.3. Discussion

There is a large number of approaches for solving the ontology alignment problem. Most of these approaches tackle the problem by computing matrices denoting the similarities of all entity pairs from two ontologies. These matrix-based approaches are prone to scalability problems and lack the possibility of having the similarity of an entity pair being dependent on other correspondences in the same alignment. Moreover, the extraction of an alignment from the matrix or matrices requires all similarities to be already computed. Hence, these approaches cannot be interrupted to retrieve intermediate alignment results. Constraint-based approaches, such as the CODI system, exist but tend to suffer from scalability problems due to their complex internal inferencing mechanism.

On the other hand, biologically-inspired optimisation techniques have recently attracted interest in the Semantic Web community and have been applied for a variety of problems, including ontology alignment. However, the existing biologically-inspired optimisation approaches for ontology alignment either impose unacceptable alignment constraints, or target only the sub-topic of finding an optimal parameter or weight configuration for an alignment algorithm. Apart from existing approaches in the actual problem domain of ontology alignment, there are plenty of applications of biologically-inspired optimisation techniques for problems that are structurally similar to ontology alignment. The problem of attribute selection for a machine learning classifier, as well as several assignment type problems are relevant examples that suggest the feasibility of biologically-inspired optimisation in the context of ontology alignment.

4. Evaluation Metrics for Ontology Alignment

In order for an algorithm to determine the optimal alignment between two ontologies, it is crucial to have a formal notion of what an optimal alignment is. This notion of optimality in turn requires any two alignments to be comparable with respect to their quality in order to make a statement such as "Alignment A is better than alignment B". Vaguely speaking, an alignment is better than another one if it better fulfils the requirements and expectations of the alignment consumer. Ideally these requirements and expectations are reflected by a set of explicitly defined quality criteria according to which an alignment can be assessed.

The flexibility of ontology languages and the various different ways ontologies are modelled in practise make it difficult to define simple alignment quality criteria according to which such an alignment evaluation can be computed. The following examples show how entirely different quality criteria are necessary to define a useful evaluation.

Example. Consider two ontologies and an alignment between them being interpreted only in terms of the logical underpinnings that determine the formal semantics of ontological entities. The alignment is considered as a set of equivalence axioms represented by the correspondences. In this strict logical sense, every alignment is optimal, which does not cause the merged ontologies augmented by the alignment to become inconsistent or contain unsatisfiable classes, since there exists at least one model that satisfies all axioms.

Example. Consider two ontologies of low expressiveness developed by the same person or company but for different purposes. The ontologies have a significant overlap, and equal entity identifiers are used in these overlapping parts. The ontology language feature of *negation*[1] is not used in either ontology, thus no inconsistency can be induced by any possible

---

[1]Negation also captures "convenience" axioms, such as disjointness, which resolve to axioms containing negation after rewriting.

alignment. In this case non-logical quality criteria have to be applied for evaluating an alignment.

Example. Consider two ontologies of low expressiveness where one is a copy of the other with annotation values of entities translated into another language. Since all other ontology modelling characteristics remain equal, exploiting natural language dictionaries is crucial for comparing alignment candidates and finding an optimal one.

Example. Consider two ontologies of low expressiveness about the same domain of interest developed in different countries with annotation values of entities in different natural languages. Since different people or companies created these ontologies, they are most likely structurally different, cover a slightly wider or narrower domain, or represent a less or more detailed formalisation of the domain. Various ontology features need to be considered in this case in order to assess the quality of an alignment, since neither logical, linguistic, lexical, or structural characteristics alone might be sufficient to reflect alignment quality.

While in the first example only semantically significant language features need to be considered in order to determine whether an alignment is optimal, the other examples require additional techniques. In the second example, no inconsistent ontology could be induced by any alignment, however, the clear overlap (in terms of shared entity identifiers) indicates that entities with the same identifier should correspond. Entirely different identifiers might be present in the third example, however, the exploitation of an appropriate dictionary clearly indicates corresponding entities. In the fourth example, several ontology features, and possibly domain-specific background knowledge, need to be exploited and combined in order to detect the best correspondences.

As the examples show, identifying universal criteria that describe the quality of alignments is a difficult problem. What is intuitively understood as a high quality alignment depends on mainly two factors:

- Use case of the alignment, *i.e.* the way an alignment is used by a semantic application.

- Elaborateness of the ontologies, *i.e.* the extent to which specific domain knowledge is made explicit in the ontology models. This extent might vary due to the use case of the ontology (where less or more details about the domain of interest might be required), and the expressiveness of the ontology language used.

It can be observed that most ontologies found on the Web are of low logical expressiveness [14]. The notion of ontologies being an *"explicit specification"* [59] is thus to a large extent being neglected by leaving a significant portion of that specification implicitly encoded in natural language annotations. For that reason several ontology alignment systems such as CTXMATCH [25] or S-Match [56] focus on the exploitation of natural language semantics [55] in order to identify relations between ontological entities.

This chapter describes various evaluation metrics for both correspondences and alignments that reflect some interesting and frequently occurring evaluation criteria, as well as means to combine them to a single evaluation score for an alignment. However, it should be noted that this selection is by no means exhaustive, *i.e.* some alignment use cases might require additional criteria to be encoded, which are not described here. In fact, the evaluation metrics proposed here are of a rather general nature in order to provide a proof-of-concept implementation of a fitness function for the optimisation algorithms presented in Chapter 5. It is not the goal in this work to provide an algorithm and according evaluation metrics for a particular alignment use case or scenario, for which specifically tailored evaluation metrics might be required.

In the following Section 4.1 various evaluation metrics are presented, which are organised in three levels. The first level, presented in Section 4.1.1 describes local evaluation metrics for correspondences without context, while Section 4.1.2 describes the second level, namely correspondence evaluation metrics that respect the alignment context for the correspondence under evaluation. Section 4.1.3 presents the third level, namely global evaluation metrics for complete alignments. Several ways of combining these evaluation metrics are presented in Section 4.2 in terms of aggregation functions.

4.1. Evaluation Metrics

Every quality criterion that is to be considered when assessing an alignment is reflected by an evaluation metric. There are metrics for evaluating individual correspondences, and metrics for evaluating alignments as a whole. Correspondence evaluation metrics are further divided into those taking into account solely information about the corresponding entities in their ontology context, and those additionally taking into account information from the alignment context, *i.e.* information derived from other correspondences present in the alignment.

4.1.1. Local Correspondence Evaluation

A correspondence between two ontological entities can be evaluated according to information about the entities and the ontologies in whose vocabularies they occur.

Definition 4.1. Let $\mathcal{O}_1$ and $\mathcal{O}_2$ be ontologies. Let $C = \langle e, f \rangle$ be a correspondence with $e \in \text{voc}_t(\mathcal{O}_1)$ and $f \in \text{voc}_t(\mathcal{O}_2)$ for some $t \in \mathcal{T}$. A *local correspondence evaluation metric* is a function $h : \mathcal{C} \to [0, 1]$ computing an evaluation score $h(C)$ for C, solely based on information attached to or derivable for e (and f respectively) in the context of $\mathcal{O}_1$ (and $\mathcal{O}_2$ respectively). An evaluation score for C reflects a similarity of e and f. To this end, an evaluation score of 1 means highest similarity, an evaluation score of 0 means lowest similarity.

Local correspondence evaluation metrics typically exploit annotations of entities. The following extraction functions provide access to the various annotation values.

Definition 4.2. Let $\mathcal{O}$ be an ontology and let $e \in \text{voc}(\mathcal{O})$ be an entity. The *identifier extractor* is defined as a function $\text{id}_{\mathcal{O}} : \text{voc}(\mathcal{O}) \to$ `String` that maps an entity to its identifier.

Example. In the case of OWL, the identifier extractor returns the IRI fragment of the entity.

Definition 4.3. Let $\mathcal{O}$ be an ontology and let $e \in \text{voc}(\mathcal{O})$ be an entity. The *label extractor* $\text{label}_{\mathcal{O}} \in \text{annot}(\mathcal{O})$ is defined as a function $\text{label}_{\mathcal{O}} : \text{voc}(\mathcal{O}) \to 2^{\texttt{String}}$ that maps an entity to its labels. If there is no label assigned to an entity, the label extractor is undefined. If there is exactly one label assigned to an entity, the label extractor delivers a singleton containing exactly this label.

Example. In the case of the ontology language OWL, the label extractor returns the set of `rdfs:label` annotation values, or is undefined if there is no such annotation.

Definition 4.4. Let $\mathcal{O}$ be an ontology and let $e \in \text{voc}(\mathcal{O})$ be an entity. The *comment extractor* $\text{comment}_{\mathcal{O}} \in \text{annot}(\mathcal{O})$ is defined as a function

$comment_{\mathcal{O}} : voc(\mathcal{O}) \rightarrow 2^{\texttt{String}}$ that maps an entity to its comments. If there is no comment assigned to an entity, the comment extractor is undefined. If there is exactly one comment assigned to an entity, the comment extractor delivers a singleton containing exactly this comment.

Example. In the case of the ontology language OWL, the comment extractor returns the `rdfs:comment` annotation values, or is undefined if there is no such annotation.

Analogously to these extractor functions there can be arbitrary other extractor functions to obtain annotations defined by specific vocabularies, such as the Dublin Core® Metadata Element Set[2] (*e.g.* `dcterms:title`), the Friend-of-a-Friend (FOAF) Vocabulary[3] (*e.g.* `foaf:name`), the Simple Knowledge Organization System (SKOS) [6] (*e.g.* `skos:prefLabel`), *etc.*, depending on their occurrence in the ontologies to be aligned.

The following paragraphs define some local correspondence evaluation metrics, where each of these metrics represents a quality criterion for correspondences.

Lexical Similarity

There are cases where it is required to compare text strings on the character level, in order to come up with a similarity for two entities. This *lexical similarity* is typically useful in alignment scenarios where different abbreviations, whitespace encoding (such as "CamelCase", dash, or underscore), or upper-/lowercase policies are applied.

Example. Considering the ontologies presented in Figure 1.1, different abbreviation schemes are used, for instance, for the entities `TechReport` and `TechnicalReport`. An example for upper-/lowercase variation are the entities `Incollection` and `InCollection`.

Several metrics have been proposed for computing a similarity (or distance) measure for two strings. A prominent measure is the *Levenshtein distance* [82], which, for two strings s_1 and s_2, denotes the minimum number of substitution, insertion, and deletion operations required to transform s_1 into s_2 (or vice versa).

Another metric for computing a string similarity has been proposed by Stoilos *et al.* [120], which is specially tailored to meet the requirements frequently faced in the context of ontology alignment. Thus it overcomes several shortcomings of other metrics, such as the Levenshtein distance, which have originally been designed for use cases other than ontology alignment. The *String Metric for Ontology Alignment* (SMOA) proposed by Stoilos *et al.* builds on the notion of common substrings and the lengths of the remaining unmatched substrings. The overall SMOA function is defined as $smoa : \texttt{String} \times \texttt{String} \to [0, 1]$. Let s_1 and s_2 be two strings. For any string s let $|s|$ denote its length. Let CS be the set of common substrings of s_1 and s_2, such that for any two substrings $cs_1, cs_2 \in CS$ holds cs_1 is not a substring of cs_2 and vice versa. Let u_1 be the string resulting from s_1 after removing all $cs \in CS$, and let u_2 be the string resulting from s_2 after removing all $cs \in CS$.

$$smoa(s_1, s_2) = c(s_1, s_2) - d(s_1, s_2) + w(s_1, s_2) \tag{4.1}$$

where

$$c(s_1, s_2) = \frac{2 \sum_{cs \in CS} |cs|}{|s_1| + |s_2|} \tag{4.2}$$

contributes positively to the similarity according to the lengths of commons substrings, and

$$d(s_1, s_2) = \frac{|u_1| \cdot |u_2|}{p + (1 - p) \cdot (|u_1| + |u_2| - |u_1| \cdot |u_2|)} \tag{4.3}$$

contributes negatively to the similarity according to the lengths of unmatched substrings. The non-negative parameter p weights the contribution of this "difference" component of the similarity due to the intuition that the difference should contribute less to the overall similarity than the commonality. $w(s_1, s_2)$ is used for "improvement of the result using the method introduced by Winkler [133]" [120].

Lexical Entity Identifier Similarity. In many cases equal identifiers are used for entities in different ontologies. Not only artificial identifiers, such as product codes, can occur, but also natural language names. The ontologies from the introductory example in Figure 1.1 use natural language names as identifiers. As this example shows, variations in spelling, formatting, and abbreviation require the application of sophisticated similarity metrics. The *lexical entity identifier similarity* denotes the similarity of two text strings that represent the identifiers of the two entities to be compared. In OWL, the identifier of an entity is an IRI, however, since typically the namespace context is not significant for similarity computation, this comparison only considers IRI fragments.

Let $e_1 \in \text{voc}(\mathcal{O}_1)$ and $e_2 \in \text{voc}(\mathcal{O}_2)$ be two entities. The *lexical entity identifier similarity* is defined as

$$h_{\text{lexIDSim}}(\langle e_1, e_2 \rangle) = smoa(\text{id}_{\mathcal{O}_1}(e_1), \text{id}_{\mathcal{O}_2}(e_2)) \tag{4.4}$$

Note that this definition uses the SMOA similarity to compute a lexical similarity. It can be replaced by any other function for string similarity, such as the Levenshtein distance.

Lexical Entity Label Similarity.   Entities can be annotated with labels. The *lexical entity label similarity* denotes the similarity of two text strings that represent the labels of the two entities to be compared.

Let $e_1 \in \text{voc}(\mathcal{O}_1)$ and $e_2 \in \text{voc}(\mathcal{O}_2)$ be two entities. The *lexical entity label similarity* is defined as

$$h_{\text{lexLabelSim}}(\langle e_1, e_2 \rangle) = smoa(\text{label}_{\mathcal{O}_1}(e_1), \text{label}_{\mathcal{O}_2}(e_2)) \tag{4.5}$$

Analogously to the lexical entity identifier similarity, the SMOA similarity can be replaced by any other function for string similarity, such as the Levenshtein distance.

Linguistic Similarity

Since ontological entities typically have real-world analogies, it is common to observe natural language labels, comments, or other annotations. If sufficient for the use case at hand[4], the semantics of an entity is often not completely explicitly formalised in the ontology, but remains intrinsic to the natural language annotations describing the entities. In those cases it is indispensable to include natural language processing techniques in order to assess correspondences.

A frequently applied technique for computing a similarity between two natural language texts proposed by Salton *et al.* is the *vector space model* [113]. In order to obtain the vector space similarity, let d and d' be two natural language texts to be compared. Let $T = \{t_1, t_2, \ldots, t_n\}$

---

[4]For example, if the ontologies are primarily used for human consumption.

be the set of distinct terms occurring in d, where n denotes the number of distinct terms in d. Analogously, let $T' = \{t_1', t_2', \ldots, t_{n'}'\}$ be the set of distinct terms occurring in d' where n' denotes the number of distinct terms in d'. For the texts d and d', a *bag of words* is defined as the union $\bar{T} = T \cup T' = \{\bar{t}_1, \bar{t}_2, \ldots, \bar{t}_m\}$. This set representation contains all terms occurring in d or d' with no duplicates and can now be denoted as a vector $\vec{T} = (\bar{t}_1, \bar{t}_2, \ldots, \bar{t}_m)$. Let $\vec{U} = (u_1, u_2, \ldots, u_m)$ be a vector representation of d, where each u_j is the number of occurrences of $\bar{t}_j$ in d. Analogously, let $\vec{U}' = (u_1', u_2', \ldots, u_m')$ be a vector representation of d', where each u_j' is the number of occurrences of $\bar{t}_j$ in d'. The *vector space similarity* of two texts d and d is the cosine of the angle ϕ between $\vec{U}$ and $\vec{U}'$

$$\cos \phi = \frac{\vec{U} \cdot \vec{U}'}{|\vec{U}| \cdot |\vec{U}'|} \tag{4.6}$$

Textual Entity Identifier Similarity. Natural language terms are frequently used as entity identifiers. This does not exclude the use of multi-word identifiers, as long as they meet the requirements for valid and unique identifiers. For instance, the identifier `Fun-And-Leisure` is a multi-word, valid IRI fragment. After removal of word separator characters and stop words, the identifier can be interpreted as natural language text and the vector space model can be applied in order to compute a similarity between two identifiers. Let $\langle e_1, e_2 \rangle$ be a correspondence between two entities $e_1 \in \mathrm{voc}(\mathcal{O}_1)$ and $e_2 \in \mathrm{voc}(\mathcal{O}_2)$ Let d_1 and d_2 be the natural language texts extracted for $\mathrm{id}_{\mathcal{O}_1}(e_1)$ and $\mathrm{id}_{\mathcal{O}_2}(e_2)$, respectively. Let $\cos \phi$ be the cosine angle for the vector representations of d_1 and d_2 as defined in Equation (4.6). The *textual entity identifier similarity* is defined as

$$h_{\mathrm{textIDSim}}(\langle e_1, e_2 \rangle) = \cos \phi \tag{4.7}$$

Textual Entity Label Similarity. Natural language terms are frequently used in entity labels. This does not exclude the use of multi-word identifiers. For instance, the label "conference proceedings editor" is a multi-word entity label. After stop word removal, the label can be interpreted as natural language text and the vector space model can be applied in order to compute a similarity between two labels. Let $\langle e_1, e_2 \rangle$ be a correspondence with $e_1 \in \mathrm{voc}(\mathcal{O}_1)$ and $e_2 \in \mathrm{voc}(\mathcal{O}_2)$. Let d_1 and d_2 be the natural language texts extracted from $\mathrm{label}_{\mathcal{O}_1}(e_1)$ and $\mathrm{label}_{\mathcal{O}_2}(e_2)$, respectively. Let $\cos \phi$ be the cosine angle for the vector representations of d_1

and d_2 as defined in Equation (4.6). The *textual entity label similarity* is defined as

$$h_{\text{textLabelSim}}(\langle e_1, e_2 \rangle) = \cos \phi \tag{4.8}$$

Entity Comment Similarity.   Comment annotations of entities may contain sentences or phrases to describe an entity in natural language. After stop word removal, the vector space model can be applied in order to compute a similarity between two comments. Let $\langle e_1, e_2 \rangle$ be a correspondence with $e_1 \in \text{voc}(\mathcal{O}_1)$ and $e_2 \in \text{voc}(\mathcal{O}_2)$. Let d_1 and d_2 be the natural language texts extracted from $\text{comment}_{\mathcal{O}_1}(e_1)$ and $\text{comment}_{\mathcal{O}_2}(e_2)$, respectively. Let $\cos \phi$ be the cosine angle for the vector representations of d_1 and d_2 as defined in Equation (4.6). The *entity comment similarity* is defined as

$$h_{\text{entityCommentSim}}(\langle e_1, e_2 \rangle) = \cos \phi \tag{4.9}$$

Virtual Entity Document Similarity.   Apart from label and comment annotations, arbitrary RDF vocabularies, such as the Simple Knowledge Organization System (SKOS) [6], the Dublin Core® Metadata Element Set[5], the Friend-of-a-Friend (FOAF) Vocabulary[6], *etc.*, are frequently used to annotate ontological entities. Additionally, any custom annotations can be defined and used in an ontology. The collection of all annotation values that are relevant for an entity can provide an accurate description of this entity, independent of the specific annotation vocabularies used. Motivated by this variety of mostly linguistic information represented by those annotation values, Qu *et al.* introduced the notion of *virtual documents* [103]. These represent a collection of relevant annotation values for an entity, which can then be used for computing a vector space similarity between any two entities.

While Qu *et al.* define virtual documents for general RDF graphs, the remainder of this paragraph defines an adaptation for more axiomatised ontology languages, such as OWL, taking into account different entity types and relations among them. To this end, the annotation values regarded relevant for an entity depends on the entity type. Let $\mathcal{O}$ be an

---

[5]`http://dublincore.org/documents/2010/10/11/dces/` (accessed March 11, 2012)

[6]`http://xmlns.com/foaf/spec/` (accessed March 11, 2012)

OWL ontology[7]. Let $e \in \text{voc}(\mathcal{O})$ be an entity of type $\tau(e) \in \mathcal{T}$, with

$$\mathcal{T} = \{\texttt{class}, \texttt{object\_property}, \texttt{data\_property}, \texttt{individual}\}$$

For any entity $e \in \text{voc}(\mathcal{O})$, let

$$
\begin{aligned}
V(e) \quad = \quad & [t_{(1,1)}, \ldots, t_{(1,n_1)}, \ldots \\
& t_{(i,1)}, \ldots, t_{(i,n_i)}, \ldots \\
& t_{(m,1)}, \ldots, t_{(m,n_m)}]
\end{aligned}
\tag{4.10}
$$

be a list of terms, where $t_{(j,1)}, \ldots, t_{(j,n_j)}$ are terms occurring in $a_j(e)$ with $a_j \in \text{annot}(\mathcal{O})$ for $1 \leq j \leq m$. Informally, if e has m annotations, then $V(e)$ is a virtual document containing all terms (including duplicates) that occur in the annotation values of the annotations of e.

The following expressions require the definition of an associative concatenation operator $\circ$ for virtual documents, such that for any two virtual documents $V = [t_1, \ldots, t_n]$ and $W = [u_1, \ldots, u_m]$, their concatenation is $V \circ W = [t_1, \ldots, t_n, u_1, \ldots, u_m]$.

Depending on $\tau(e)$, the set of *relevant* entities $rel(e)$ for collecting annotation values is defined as follows:

- $\tau(e) = \texttt{class}$:

  Let $\text{sup}_{\mathcal{O}}(e)$ be the set of direct superclasses of e in $\mathcal{O}$. Let $\text{sub}_{\mathcal{O}}(e)$ be the set of direct subclasses of e in $\mathcal{O}$. Let $\text{dom}_{\mathcal{O}}(e)$ be the set of entities in the description of the properties in $\mathcal{O}$ for which e occurs in the domain restriction. Let $\text{range}_{\mathcal{O}}(e)$ be the set of entities in the description of the object properties in $\mathcal{O}$ for which e occurs in the range restriction. Let $\text{ind}_{\mathcal{O}}(e)$ be the set of individuals asserted to e in $\mathcal{O}$. Then the virtual document for $rel(e)$ is defined as

$$
\begin{aligned}
V(rel(e)) \quad = \quad & V(e) \circ \\
& V(\text{sup}_{\mathcal{O}}(e)) \circ V(\text{sub}_{\mathcal{O}}(e)) \circ \\
& V(\text{dom}_{\mathcal{O}}(e)) \circ V(\text{range}_{\mathcal{O}}(e)) \circ \\
& V(\text{ind}_{\mathcal{O}}(e))
\end{aligned}
\tag{4.11}
$$

- $\tau(e) = \texttt{object\_property}$:

  Let $\text{sup}_{\mathcal{O}}(e)$ be the set of direct object superproperties of e in $\mathcal{O}$.

---

[7]Depending on the ontology language, the notion of relevance for annotation values with respect to an entity has to be determined accordingly.

Let $\mathrm{sub}_{\mathcal{O}}(e)$ be the set of direct object subproperties of e in $\mathcal{O}$. Let $\mathrm{dom}_{\mathcal{O}}(e)$ be the set of entities occurring in the class descriptions of the domain restriction of e in $\mathcal{O}$. Let $\mathrm{range}_{\mathcal{O}}(e)$ be the set of entities occurring in the class description of the range of e in $\mathcal{O}$. Let $\mathrm{ind}_{\mathcal{O}}(e)$ be the set of individuals asserted to e in $\mathcal{O}$. Then the virtual document for $rel(e)$ is defined as

$$
\begin{aligned}
V(rel(e)) \;=\; & V(e) \circ \\
& V(\mathrm{sup}_{\mathcal{O}}(e)) \circ V(\mathrm{sub}_{\mathcal{O}}(e)) \circ \\
& V(\mathrm{dom}_{\mathcal{O}}(e)) \circ V(\mathrm{range}_{\mathcal{O}}(e)) \circ \\
& V(\mathrm{ind}_{\mathcal{O}}(e))
\end{aligned}
\tag{4.12}
$$

- $\tau(e) = \texttt{data\_property}$:
  Let $\mathrm{sup}_{\mathcal{O}}(e)$ be the set of direct data superproperties of e in $\mathcal{O}$. Let $\mathrm{sub}_{\mathcal{O}}(e)$ be the set of direct data subproperties of e in $\mathcal{O}$. Let $\mathrm{dom}_{\mathcal{O}}(e)$ be the set of entities occurring in the class descriptions of the domain restriction of e in $\mathcal{O}$. Let $\mathrm{ind}_{\mathcal{O}}(e)$ be the set of individuals asserted to e in $\mathcal{O}$. The virtual document for $rel(e)$ is defined as

$$
\begin{aligned}
V(rel(e)) \;=\; & V(e) \circ \\
& V(\mathrm{sup}_{\mathcal{O}}(e)) \circ V(\mathrm{sub}_{\mathcal{O}}(e)) \circ \\
& V(\mathrm{dom}_{\mathcal{O}}(e)) \circ \\
& V(\mathrm{ind}_{\mathcal{O}}(e))
\end{aligned}
\tag{4.13}
$$

- $\tau(e) = \texttt{individual}$:
  Let $\mathrm{cls}_{\mathcal{O}}(e)$ be the set of classes to which e is asserted in $\mathcal{O}$. Let $\mathrm{sbj}_{\mathcal{O}}(e)$ be the set of object or data properties in $\mathcal{O}$ for which e is subject. Let $\mathrm{obj}_{\mathcal{O}}(e)$ be the set of object properties in $\mathcal{O}$ for which e is object. Then the virtual document for $rel(e)$ is defined as

$$
\begin{aligned}
V(rel(e)) \;=\; & V(e) \circ \\
& V(\mathrm{cls}_{\mathcal{O}}(e)) \circ \\
& V(\mathrm{sbj}_{\mathcal{O}}(e)) \circ \\
& V(\mathrm{obj}_{\mathcal{O}}(e))
\end{aligned}
\tag{4.14}
$$

Let $e_1 \in \mathrm{voc}(\mathcal{O}_1)$ and $e_2 \in \mathrm{voc}(\mathcal{O}_2)$ be entities. Then $d_1 = V(rel(e_1))$ and $d_2 = V(rel(e_2))$ are the *virtual documents* for e_1 and e_2, respectively.

Let $\cos\phi$ be the cosine angle for the vector representations of d_1 and d_2 as defined in Equation (4.6). Then the *virtual entity document similarity* is defined as

$$h_{\mathrm{entityVDSim}}(\langle e_1, e_2\rangle) = \cos\phi \tag{4.15}$$

4.1.2. Contextual Correspondence Evaluation

A correspondence between two ontological entities can be evaluated according to information known for these entities in their ontology context, as well as in the context of the alignment the correspondence under consideration is an element of.

Definition 4.5. Let $\mathcal{O}_1$ and $\mathcal{O}_2$ be ontologies. Let A be an alignment between $\mathcal{O}_1$ and $\mathcal{O}_2$, and let $C = \langle e, f\rangle \in A$ be a correspondence in A. A *contextual correspondence evaluation metric* is a function $h^A : \mathcal{C} \to [0, 1]$ computing an evaluation score $h^A(C)$ for C in the alignment context A based on the information attached to or derivable for e (and f respectively) in the context of $\mathcal{O}_1$ (and $\mathcal{O}_2$ respectively) and A. An evaluation score for C reflects a similarity of e and f. Thereby, an evaluation score of 1 means highest similarity, and an evaluation score of 0 means lowest similarity.

Contextual correspondence evaluation metrics exploit the mere presence, or the confidence of other correspondences in the same alignment context. These other correspondences refer to entities, whose relation to the entities of the correspondence under evaluation is important. For a given entity, the following extraction functions provide access to other entities in the same ontology.

Definition 4.6. Let $\mathcal{O}$ be an ontology. Let $e \in \mathrm{voc}(\mathcal{O})$ be an entity with $\tau(e)$ being an entity type that allows for a subsumption relation $\sqsubseteq$. A *neighbour* of e with respect to $\sqsubseteq$ is an entity $f \in \mathrm{voc}(\mathcal{O})$ with $\tau(f) = \tau(e)$, such that $f \sqsubseteq e$ ($e \sqsubseteq f$) and there is no f', such that $f \sqsubseteq f' \sqsubseteq e$ (or $e \sqsubseteq f' \sqsubseteq f$, respectively).

If $\sqsubseteq$ denotes the explicit subsumption relation, *i.e.* all subsumption axioms are explicitly stated in the ontology, every neighbour of an entity e with respect to $\sqsubseteq$ is called *explicit neighbour*.

Definition 4.7. Let $\mathcal{O}$ be an ontology and let $e \in \mathrm{voc}(\mathcal{O})$ be an entity with $\tau(e)$ being an entity type that allows for a subsumption relation. The *superentity extractor* is defined as a function $\mathrm{sup}_{\mathcal{O}}(e) : \mathrm{voc}(\mathcal{O}) \to 2^{\mathrm{voc}(\mathcal{O})}$, where $\mathrm{sup}_{\mathcal{O}}(e) = \{f \in \mathrm{voc}(\mathcal{O}) \mid \tau(e) = \tau(f),\ e \sqsubseteq f\}$ is the set of superentities of e in $\mathcal{O}$.

Definition 4.8. Let $\mathcal{O}$ be an ontology and let $e \in \mathrm{voc}(\mathcal{O})$ be an entity with $\tau(e)$ being an entity type that allows for a subsumption relation. The *subentity extractor* is defined as a function $\mathrm{sub}_{\mathcal{O}}(e) : \mathrm{voc}(\mathcal{O}) \rightarrow 2^{\mathrm{voc}(\mathcal{O})}$, where $\mathrm{sub}_{\mathcal{O}}(e) = \{f \in \mathrm{voc}(\mathcal{O}) \mid \tau(e) = \tau(f),\ f \sqsubseteq e\}$ is the set of subentities of e in $\mathcal{O}$.

Example. In the case of OWL, the superentity extractor would return all direct superclasses for any entity of type `class`, all direct object superproperties for any entity of type `object_property`, and all direct data superproperties for any entity of type `data_property`. The subentity extractor would analogously return the according subentities [92].

The following paragraphs define a number of contextual correspondence evaluation metrics. Each metric represents a quality criterion for correspondences.

Hierarchy Similarity

For entities of certain types a subsumption relation can be expressed, forming a subsumption hierarchy.

The *hierarchy similarity* denotes the similarity of two entities based on the presence of correspondences between neighbour entities in the alignment context. It is defined as

$$h^{A}_{\mathrm{hierarchy}}(\langle e, f \rangle) = \omega_{\mathrm{sup}} \frac{|\{\langle e', f' \rangle \in A \mid e' \in \mathrm{sup}_{\mathcal{O}_1}(e),\ f' \in \mathrm{sup}_{\mathcal{O}_2}(f)\}|}{\min\{\mathrm{sup}_{\mathcal{O}_1}(e),\ \mathrm{sup}_{\mathcal{O}_2}(f)\}} +$$
$$\omega_{\mathrm{sub}} \frac{|\{\langle e', f' \rangle \in A \mid e' \in \mathrm{sub}_{\mathcal{O}_1}(e),\ f' \in \mathrm{sub}_{\mathcal{O}_2}(f)\}|}{\min\{\mathrm{sub}_{\mathcal{O}_1}(e),\ \mathrm{sub}_{\mathcal{O}_2}(f)\}}$$

$$(4.16)$$

where

$$\omega_{\mathrm{sup}} = \frac{\min\{\mathrm{sup}_{\mathcal{O}_1}(e),\ \mathrm{sup}_{\mathcal{O}_2}(f)\}}{\min\{\mathrm{sup}_{\mathcal{O}_1}(e),\ \mathrm{sup}_{\mathcal{O}_2}(f)\} + \min\{\mathrm{sub}_{\mathcal{O}_1}(e),\ \mathrm{sub}_{\mathcal{O}_2}(f)\}} \qquad (4.17)$$

and $\omega_{\mathrm{sub}} = 1 - \omega_{\mathrm{sup}}$. Informally, the hierarchy similarity composes of two components for superentities and subentities, respectively. Each component denotes the fraction of potential correspondences of super- or subentities that are actually contained in the alignment. The contribution of each component is weighted according to a heuristic indicating the importance of super- and subentities. The idea behind looking at the smaller set of super- and subentities is that this size determines the maximum number of possible correspondences between super- or subentities. If, for instance, there is larger number of potential subentity correspondences than superentity correspondences, the subentity component should be weighted higher.

Example. Given the two ontologies from Figure 1.1, denoted as $\mathcal{O}_{1.1a}$ and $\mathcal{O}_{1.1b}$, let $C = \langle \texttt{Entry}, \texttt{Publication} \rangle \in A$ be a correspondence. The numbers of super- and subentities for the two entities of C are

$$
\begin{aligned}
\text{sup}_{\mathcal{O}_{1.1a}}(\texttt{Entry}) &= 1 \\
\text{sub}_{\mathcal{O}_{1.1a}}(\texttt{Entry}) &= 14 \\
\text{sup}_{\mathcal{O}_{1.1b}}(\texttt{Publication}) &= 1 \\
\text{sub}_{\mathcal{O}_{1.1b}}(\texttt{Publication}) &= 12
\end{aligned}
$$

According the the weighting heuristic, $\omega_{\text{sup}} = 1/13$ and $\omega_{\text{sub}} = 12/13$. From this example one can see that the heuristic reflects the intuition that a large number of correspondences between subentities should be considered as more important than the single possible correspondence between the superentities.

Suppose A contains the following correspondences in addition to C:

$\langle \texttt{Thing}, \texttt{Thing} \rangle, \langle \texttt{Article}, \texttt{Article} \rangle, \langle \texttt{Book}, \texttt{Book} \rangle,$
$\langle \texttt{Booklet}, \texttt{Booklet} \rangle, \langle \texttt{Manual}, \texttt{Manual} \rangle, \langle \texttt{Misc}, \texttt{Misc} \rangle,$
$\langle \texttt{Unpublished}, \texttt{Unpublished} \rangle, \langle \texttt{Proceedings}, \texttt{Proceedings} \rangle.$

These are 7 out of 12 possible correspondences between subentities and the only one possible correspondence between superentities. Using the weights determined above, the hierarchy similarity of C computes as

$$
h_{\text{hierarchy}}^{A}(C) \quad = \quad \frac{1}{13} \cdot \frac{1}{1} + \frac{12}{13} \cdot \frac{7}{12} \quad \approx \quad 0.615
$$

Hierarchy Propagation Similarity

Similar to the hierarchy similarity is the *hierarchy propagation similarity*, however, this one does not account for the fraction of potential correspondences of super- or subentities. Instead, it propagates the similarities from correspondences of super- and subentities to the correspondence under evaluation. The computation of this similarity can be seen as a simplified, non-iterative version of the similarity flooding algorithm[8] introduced by Melnik *et al.* [90]. It is following the intention that if two nodes of two different graphs are considered similar, then the "neighbours" of those nodes can be considered somewhat similar, too. The similarity thus propagates to neighbouring nodes. Deviating from the authors' original work, when transferring their generic graph matching algorithm to the more specific problem of ontology alignment, nodes, edges, and their labels must be interpreted in ontological terms. What can be straightforward in RDF graphs becomes more challenging in RDFS- and OWL-based ontologies, where classes and properties are first class entities with specific axiomatic relationships, such as subsumption, equivalence, or domain and range restrictions.

Compared to similarity flooding the *hierarchy propagation similarity* here is simplified, since only the subsumption hierarchy is considered when determining "neighbours", disregarding arbitrary RDF triple "edges". Furthermore, only a single correspondence is evaluated with similarities propagated from those "neighbouring" entities. Moreover, the *hierarchy propagation similarity* is non-iterative, since only a snapshot evaluation is computed based on similarities of neighbouring entities, which might not be stable themselves.

Let $\mathcal{O}_1$ and $\mathcal{O}_2$ be two ontologies, and let $e_1 \in \text{voc}(\mathcal{O}_1)$ and $e_2 \in \text{voc}(\mathcal{O}_2)$

---

[8]In their original work, Melnik *et al.* allow arbitrary labelled edges, where similarities only propagate along edges of the same label.

Technically, for two directed graphs $G_1 = (V_1, E_1)$ and $G_2 = (V_2, E_2)$ with labelled nodes and edges, a *pairwise connectivity graph* $G_{pc} = (V_{pc}, E_{pc})$ is constructed, such that $V_{pc} \subseteq V_1 \times V_2$ and $E_{pc} \subseteq V_{pc} \times V_{pc}$. In G_{pc} every node $v = (v_1 \in V_1, v_2 \in V_2)$ is assigned an initial similarity, *i.e.* the initial similarity between $v_1 \in V_1$ and $v_2 \in V_2$. Furthermore, two nodes $v = (v_1 \in V_1, v_2 \in V_1)$ and $w = (w_1 \in V_1, w_2 \in V_2)$ are connected via an edge $e \in E_{pc}$ iff there are edges $e_1 = (v_1, w_1) \in E_1$ and $e_2 = (v_2, w_2) \in E_2$, and e_1 and e_2 have the same label. In a next step, from the *pairwise connectivity graph* an *induced propagation graph* $G_{ip} = (V_{ip}, E_{ip})$ is constructed, such that $V_{ip} = V_{pc}$ and $E_{ip} = E_{pc} \cup \{(w, v) : (v, w) \in E_{pc}\}$. Thus G_{ip} contains all edges from G_{pc} and their inverse edges. Additionally, each edge $e \in E_{ip}$ is assigned a *propagation coefficient* $w \in [0, 1]$ denoting the contribution of a node's similarity to the similarity of its neighbour connected via e.

Based on this G_{ip} data structure an iterative fixpoint computation leads to a new assignment of similarity measures to pairs of nodes from G_1 and G_2. The authors point out that there are several ways of assigning the propagation coefficients and calculate updated similarities in each iteration step. Empirical experiments revealed insights into the behaviour of different propagation coefficients and fixpoint computation formulae. Further, the experiments demonstrated that the initial similarities assigned to node pairs have less important impact on the resulting similarities than expected.

be entities with $\tau(e_1) = \tau(e_2)$ being an entity type that allows a subsumption relation. Let

$$
\begin{aligned}
\mathcal{C}_{\text{sup}} &= \{\langle f_1, f_2 \rangle \in A \mid f_1 \in \sup_{\mathcal{O}_1}(e_1) \text{ and } f_2 \in \sup_{\mathcal{O}_2}(e_2)\} &&(4.18) \\
\mathcal{C}_{\text{sub}} &= \{\langle f_1, f_2 \rangle \in A \mid f_1 \in \text{sub}_{\mathcal{O}_1}(e_1) \text{ and } f_2 \in \text{sub}_{\mathcal{O}_2}(e_2)\} &&(4.19)
\end{aligned}
$$

be the sets of correspondences of the super- and subentities, which are also present in the alignment context. Let

$$
\mathcal{C}_{\text{neighbour}} = \mathcal{C}_{\text{sup}} \cup \mathcal{C}_{\text{sub}} \tag{4.20}
$$

be the set of neighbouring correspondences. The *hierarchy propagation similarity* is defined as

$$
h^A_{\text{hierarchyProp}}(C) = \begin{cases} \dfrac{\sum_{C' \in \mathcal{C}_{\text{neighbour}}} \iota(C')}{|\mathcal{C}_{\text{neighbour}}|} & \text{if } |\mathcal{C}_{\text{neighbour}}| \neq 0 \\ \text{undefined} & \text{otherwise} \end{cases} \tag{4.21}
$$

Example. Given the two ontologies from Figure 1.1, denoted as $\mathcal{O}_{1.1a}$ and $\mathcal{O}_{1.1b}$, let A be an alignment containing the following correspondences:

$$
\begin{aligned}
C_0 &= \langle \texttt{Entity, Publication} \rangle \\
C_1 &= \langle \texttt{Thing, Thing} \rangle \\
C_2 &= \langle \texttt{Article, Article} \rangle \\
C_3 &= \langle \texttt{Conference, Event} \rangle \\
C_4 &= \langle \texttt{Booklet, Booklet} \rangle
\end{aligned}
$$

Let the confidences of these correspondences be as follows:

$$
\begin{aligned}
\iota(C_0) &= 0.0 \\
\iota(C_1) &= 0.8 \\
\iota(C_2) &= 0.9 \\
\iota(C_3) &= 0.4 \\
\iota(C_4) &= 0.8
\end{aligned}
$$

Let C_0 be the correspondence under evaluation. With $\texttt{Thing}$ being a superentity of $\texttt{Entry}$ in $\mathcal{O}_{1.1a}$ as well as $\texttt{Thing}$ being a superentity of $\texttt{Publication}$ in $\mathcal{O}_{1.1b}$, let

$$
\mathcal{C}_{\text{sup}} = \{C_1\} = \{\langle \texttt{Thing, Thing} \rangle\}
$$

Analogously, with `Article` and `Booklet` being subentities of `Entry` in $\mathcal{O}_{1.1a}$ and `Article` and `Booklet` being subentities of `Publication` in $\mathcal{O}_{1.1b}$, let

$$\mathcal{C}_{\text{sub}} = \{C_2, C_4\} = \{\langle \texttt{Article}, \texttt{Article} \rangle, \langle \texttt{Booklet}, \texttt{Booklet} \rangle\}$$

This results in $\mathcal{C}_{\text{neighbour}} = \{C_1, C_2, C_4\}$. Using the confidence values of these correspondences, the hierarchy propagation similarity becomes

$$h^A_{\text{hierarchyProp}}(C) \quad = \quad \frac{0.8 + 0.9 + 0.8}{3} \quad \approx \quad 0.83$$

Property Domain/Range Similarity

Axioms in ontologies can express restrictions about the domain or range of a property, provided that the ontology language supports those sorts of restrictions and entity types. For instance, in OWL, domains for entities of type `object_property` and `data_property`, as well as ranges for entities of type `object_property` can be restricted to satisfy a particular `class` expression. Since this similarity is only applicable if the underlying ontology language supports the entity types and axioms described above, the remainder of this paragraph is based on the assumption that such an ontology language is used.

The *property domain/range similarity* is a similarity metric for properties, based on the similarity of (or mere presence of correspondences between) classes, to which their domains/ranges are restricted. In other words, two properties are more likely to be similar if they have their domains (or ranges) restricted to classes, which are already known to be similar (in the alignment context).

Let $\mathcal{O}_1$ and $\mathcal{O}_2$ be two ontologies, and let $e_1 \in \text{voc}_t(\mathcal{O}_1)$ and $e_2 \in \text{voc}_t(\mathcal{O}_2)$ be entities with t being a property-like entity type that can have domain and range restrictions. Let D_1 and D_2 be the sets of atomic domain classes of e_1 and e_2, respectively, where all entities in D_1 and D_2 have the same entity type. Let R_1 and R_2 be the sets of range classes of e_1 and e_2, respectively, where all entities in R_1 and R_2 have the same entity type. Let $C = \langle e_1, e_2 \rangle$ be a correspondence and let A be the alignment context in which it shall be evaluated. Let

$$\mathcal{C}_D \quad = \quad \{\langle f_1, f_2 \rangle \in A \mid f_1 \in D_1 \text{ and } f_2 \in D_2\} \tag{4.22}$$

$$\mathcal{C}_R \quad = \quad \{\langle f_1, f_2 \rangle \in A \mid f_1 \in R_1 \text{ and } f_2 \in R_2\} \tag{4.23}$$

be the sets of correspondences of the domain and range classes, which are also present in the alignment context. The similarity component considering domain class correspondences is only defined, if $\min\{|D_1|, |D_2|\} \neq 0$, *i.e.* if both e_1 and e_2 have at least one atomic domain class. In the following let $\min\{|D_1|, |D_2|\} > 0$. The derived domain class similarity is

$$d_{\text{der}} = \begin{cases} \frac{\sum_{C \in \mathcal{C}_D} \iota(C)}{|\mathcal{C}_D|} & \text{if } |\mathcal{C}_D| \neq 0 \\ 1 & \text{otherwise} \end{cases} \qquad (4.24)$$

which averages the confidences of corresponding domain classes, if there are any[9]. In case the number of domain class correspondences present in A is close to the maximum number possible (which is the smaller number of domain classes of e_1 or e_2, respectively) this is also an indicator of similarity between e_1 and e_2. If one of e_1 and e_2 does not have any domain classes, this similarity component is undefined.

$$d_{\text{num}} = \frac{|\mathcal{C}_D|}{\min\{|D_1|, |D_2|\}} \qquad (4.25)$$

In order to also account for the number of potential domain class correspondences, regardless of their presence in A, another similarity can be approximated by

$$d_{\text{pot}} = 1 - \frac{1}{\min\{|D_1|, |D_2|\} + 1} \qquad (4.26)$$

The similarity derived from the domain classes of e_1 and e_2 now computes as

$$d = \omega_{\text{der}} d_{\text{der}} + \omega_{\text{num}} d_{\text{num}} + \omega_{\text{pot}} d_{\text{pot}} \qquad (4.27)$$

where ω_{der}, ω_{num}, and ω_{pot} are weighting factors in order to account for the influence of each similarity indicator, where $\omega_{\text{der}} + \omega_{\text{num}} + \omega_{\text{pot}} = 1$.

The values for range class similarities are computed analogously. The similarity component considering range class correspondences is only defined, if $\min\{|R_1|, |R_2|\} \neq 0$, *i.e.* if both e_1 and e_2 have at least one atomic

---

[9]This derived domain class similarity is an analogous definition of the hierarchy propagation similarity for subsumption hierarchies described earlier.

range class restriction. In the following let $\min\{|R_1|, |R_2|\} > 0$.

$$r_{\mathrm{der}} = \begin{cases} \frac{\sum_{C \in \mathcal{C}_R} \iota(C)}{|\mathcal{C}_R|} & \text{if } |\mathcal{C}_R| \neq 0 \\ 1 & \text{otherwise} \end{cases} \tag{4.28}$$

$$r_{\mathrm{num}} = \frac{|\mathcal{C}_R|}{\min\{|R_1|, |R_2|\}} \tag{4.29}$$

$$r_{\mathrm{pot}} = 1 - \frac{1}{\min\{|R_1|, |R_2|\} + 1} \tag{4.30}$$

and accordingly

$$r = \omega_{\mathrm{der}} r_{\mathrm{der}} + \omega_{\mathrm{num}} r_{\mathrm{num}} + \omega_{\mathrm{pot}} r_{\mathrm{pot}} \tag{4.31}$$

The total *property domain/range similarity* computes as

$$h^A_{\mathrm{propDRClass}}(\langle e_1, e_2 \rangle) = \begin{cases} \text{undefined} & \text{if both } d \text{ and } r \text{ are undefined} \\ \text{sigmoid}\,(d) & \text{if } r \text{ is undefined} \\ \text{sigmoid}\,(r) & \text{if } d \text{ is undefined} \\ \text{sigmoid}\left(\frac{d+r}{2}\right) & \text{otherwise} \end{cases} \tag{4.32}$$

where $\text{sigmoid} : \mathbb{R} \to (0, 1)$ is a weighting function defined as

$$\text{sigmoid}(x) = \frac{1}{1 + e^{-10(x-0.5)}} \tag{4.33}$$

and illustrated in the plot in Figure 4.1. The reason for using this weighting function is to emphasise the similarity scores provided by this metric close to the co-domain of $h^A_{\mathrm{propDRClass}}$. This causes the evaluation scores to be more discriminative and results in a higher significance of this metric when computing an aggregated evaluation score for a correspondence.

Class as Domain/Range Similarity

Similar to the property domain/range similarity, this similarity is only applicable if the underlying ontology language supports the entity types and axioms that allow for domain and range restrictions, The remainder of this paragraph is based on the assumption that such an ontology language is used.

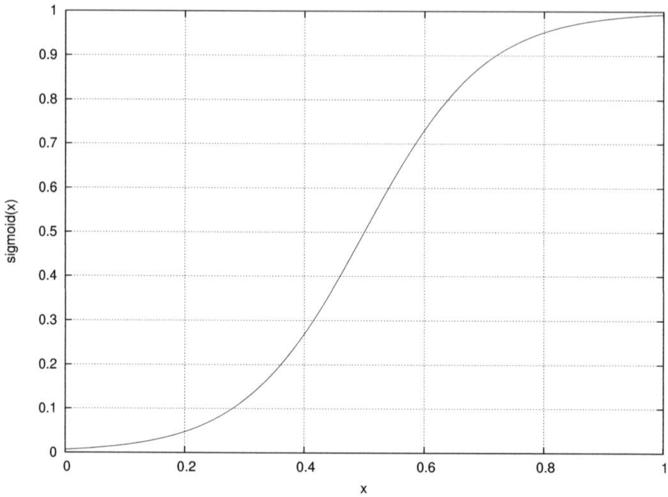

Figure 4.1.: sigmoid function $\frac{1}{1+e^{-10(x-0.5)}}$ in the relevant function domain $[0,\ 1]$.

The *class as domain/range similarity* is a similarity metric for classes that is based on the similarity of (or mere presence of correspondences between) properties with their domain/range restricted to the classes under consideration. In other words, two classes are more likely to be similar if they are both in the domain (or range) restrictions of two properties, which are already known to be similar (in the alignment context).

The following definitions are analogous to the ones defined earlier for the property domain/range similarity and follow the same reasoning.

Let $\mathcal{O}_1$ and $\mathcal{O}_2$ be two ontologies, and let $e_1 \in \mathrm{voc}_t(\mathcal{O}_1)$ and $e_2 \in \mathrm{voc}_t(\mathcal{O}_2)$ be entities with t being a class-like entity that can occur as domain or range restriction. Let P_1 and P_2 be the sets of properties that have e_1 and e_2 as domain restriction, respectively, where all entities in P_1 and P_2 have the same entity type. Let Q_1 and Q_2 be the sets of properties that have e_1 and e_2 as range restriction, respectively, where all entities in Q_1 and Q_2 have the same entity type. Let $C = \langle e_1,\ e_2 \rangle$ be a correspondence and let A be the alignment context in which it shall be evaluated. Let

$$\mathcal{C}_D = \{\langle f_1,\ f_2 \rangle \in A \mid f_1 \in P_1 \text{ and } f_2 \in P_2\} \tag{4.34}$$

$$\mathcal{C}_R = \{\langle f_1,\ f_2 \rangle \in A \mid f_1 \in Q_1 \text{ and } f_2 \in Q_2\} \tag{4.35}$$

be the sets of property correspondences in the alignment context, for which the following holds: every element in $\mathcal{C}_D$ is a correspondence of two properties, which have their domains restricted to e_1 and e_2, respectively. Analogously, every element in $\mathcal{C}_R$ is a correspondence of two properties, which have their ranges restricted to e_1 and e_2, respectively.

Given $\mathcal{C}_D$ and $\mathcal{C}_R$, the derived domain and range similarities d_{der} and r_{der} are defined exactly as in Equations (4.24) and (4.28). The other similarity components d_{num}, r_{num}, d_{pot}, and r_{pot} are defined analogously as in Equations (4.25), (4.29), (4.26), and (4.30), respectively:

$$d_{\mathrm{num}} = \begin{cases} \frac{|\mathcal{C}_D|}{\min\{|P_1|, |P_2|\}} & \text{if } \min\{|P_1|, |P_2|\} \neq 0 \\ 1 & \text{otherwise} \end{cases} \tag{4.36}$$

$$r_{\mathrm{num}} = \begin{cases} \frac{|\mathcal{C}_R|}{\min\{|Q_1|, |Q_2|\}} & \text{if } \min\{|Q_1|, |Q_2|\} \neq 0 \\ 1 & \text{otherwise} \end{cases} \tag{4.37}$$

$$d_{\mathrm{pot}} = 1 - \frac{1}{\min\{|P_1|, |P_2|\} + 1} \tag{4.38}$$

$$r_{\mathrm{pot}} = 1 - \frac{1}{\min\{|Q_1|, |Q_2|\} + 1} \tag{4.39}$$

The aggregation of similarity components from domain and range point of view is again exactly as in Equations (4.27) and (4.31), with the total *class as domain/range similarity* being computed as

$$h^A_{\mathrm{classDRProp}}(\langle e_1, e_2 \rangle) = \begin{cases} \text{undefined} & \text{if both } d \text{ and } r \text{ are undefined} \\ \text{sigmoid}(d) & \text{if } r \text{ is undefined} \\ \text{sigmoid}(r) & \text{if } d \text{ is undefined} \\ \text{sigmoid}\left(\frac{d+r}{2}\right) & \text{otherwise} \end{cases} \tag{4.40}$$

Criss-cross Correspondence

Correspondences between entities of types that allow for a subsumption relation can appear in a constellation where they "cross" each other. A correspondence $\langle e_1, e_2 \rangle$ is crossed by another correspondence $\langle f_1, f_2 \rangle$, iff f_1 is higher in the subsumption hierarchy than e_1 in $\mathcal{O}_1$, and f_2 is lower in the subsumption hierarchy than e_2 in $\mathcal{O}_2$, or vice versa. If correspondences are interpreted in a way that they induce axioms according to Definition 2.10, such a crossing would cause the hierarchies between the entities of any two crossing correspondences to collapse, which is not desired.

Via the *criss-cross correspondence evaluation metric* correspondences that participate in a crossing are penalised. It is a binary metric that scores 0 if the correspondence under evaluation is crossing at least one other correspondence, and 1 otherwise. The metric is formally defined as

$$
h^A_{\text{crissCross}}(\langle e_1, e_2 \rangle) = \begin{cases} 0 & \text{if } \exists \langle f_1, f_2 \rangle \in A \text{ such that} \\ & \quad f_1 \in \text{sub}_{\mathcal{O}_1}(e_1) \wedge f_2 \in \text{sup}_{\mathcal{O}_2}(e_2) \text{ or} \\ & \quad f_1 \in \text{sup}_{\mathcal{O}_1}(e_1) \wedge f_2 \in \text{sub}_{\mathcal{O}_2}(e_2) \\ 1 & \text{otherwise} \end{cases} \tag{4.41}
$$

Explanation-based Evaluation

If correspondences of an alignment are interpreted as equivalence axioms according to Definition 2.10, the merged ontology based on this alignment might contain unsatisfiable classes. Intuitively, if an axiom induced by a correspondence contributes to the unsatisfiability of a class, the correspondence should get a low evaluation score.

Let U be the set of unsatisfiable classes in $\mathcal{O}_1 \cup \mathcal{O}_2$. Let U_A be the set of unsatisfiable classes in $\mathcal{O}_A$. Let $\Delta U = U_A \backslash U$ be the set of classes in $\mathcal{O}_1 \cup \mathcal{O}_2$ that became unsatisfiable by adding axioms[10] induced by the alignment A. For each unsatisfiable class $u \in \Delta U$ let E be the set of axioms explaining the unsatisfiability of u by means of the "black box simple expand-shrink strategy" described by Kalyanpur [72]. Let a_C be the ontology axiom induced by C, according to Definition 2.10. The *explanation-based correspondence evaluation metric* is defined as

$$
h^A_{\text{explanation}}(C) = \begin{cases} 0 & \text{if } a_C \in E \\ 1 & \text{otherwise} \end{cases} \tag{4.42}
$$

4.1.3. Alignment Level Evaluation

A naive way of computing an evaluation score for an alignment is to aggregate the evaluation scores for each correspondence in the alignment.

---

[10]Note that for description logic based ontologies adding axioms induced by the alignment can never cause previously unsatisfiable classes to become satisfiable, since description logics are monotonic.

However, there are alignment quality criteria that cannot be measured on the level of individual correspondences, *e.g.* the alignment size.

Example. Consider the overall alignment evaluation being an aggregation of all individual correspondence evaluations. Then an alignment between two ontologies containing 10 very good correspondences and 2 bad correspondences would gain a worse evaluation than an alignment of the same ontologies that contains only 1 very good correspondence. This is due to the fact that the alignment size is not honoured.

Definition 4.9. Let $\mathcal{O}_1$ and $\mathcal{O}_2$ be ontologies. Let A be an alignment between $\mathcal{O}_1$ and $\mathcal{O}_2$. An *alignment evaluation metric* is a function $H : A \to [0, 1]$ that computes an evaluation score $H(A)$ for A. Thereby, an evaluation score of 1 means best evaluation, and an evaluation score of 0 means worst evaluation.

The following paragraphs define some alignment evaluation metrics. Each metric represents a quality criterion for alignments.

Correspondence Contribution

The straightforward determination of an alignment evaluation is to compute an aggregated score from all individual correspondence evaluations. For all $C \in A$ let $\iota(C)$ reflect the evaluation of C with respect to the various metrics presented in Sections 4.1.1 and 4.1.2. The *correspondence contribution alignment evaluation metric* is defined as

$$H_{\text{corrContrib}}(A) = \frac{1}{|A|} \sum_{C \in A} \iota(C) \tag{4.43}$$

Alignment Size

Ontologies often do not have a complete overlap, *i.e.* not every entity of either ontology has to participate in the alignment. However, it is desirable to identify the maximum overlap between ontologies, *i.e.* to prefer alignments of larger size to smaller ones. Without loss of generality[11] let

---

[11]In the case where one ontology does not contain any entities of *some* type t, the summand for t in Equation (4.44) is omitted and the normalisation is computed by dividing by $|\mathcal{T} \setminus \{t\}|$, instead of $|\mathcal{T}|$. In the case where for *all* entity types one of the ontologies does not contain any entities, no alignment is possible and the *alignment size evaluation metric* is undefined.

$\min\{\natural_t \mathcal{O}_1, \natural_t \mathcal{O}_2\} > 0, \ \forall t \in \mathcal{T}$. The *alignment size evaluation metric* is consequently defined as

$$H_{\text{size}}(A) = \frac{1}{|\mathcal{T}|} \sum_{t \in \mathcal{T}} \frac{|\{\langle e, f \rangle \in A \mid \tau(e) = \tau(f) = t\}|}{\min\{\natural_t \mathcal{O}_1, \natural_t \mathcal{O}_2\}} \tag{4.44}$$

Alignment Consistency

If correspondences of an alignment are interpreted as equivalence axioms according to Definition 2.10, the merged ontology based on this alignment might become inconsistent. Whether or not an alignment induces an inconsistency can be used as a global alignment evaluation metric. Let $\mathcal{O}_A$ be the merged ontology based on A according to Definition 2.11. The *alignment consistency evaluation metric* is defined as

$$H_{\text{consist}}(A) = \begin{cases} 1 & \text{if } \mathcal{O}_A \text{ is consistent} \\ 0 & \text{if } \mathcal{O}_A \text{ is inconsistent} \\ \text{undefined} & \text{if } \mathcal{O}_1 \text{ or } \mathcal{O}_2 \text{ is inconsistent} \end{cases} \tag{4.45}$$

Alignment Coherency

If correspondences of an alignment are interpreted as equivalence axioms according to Definition 2.10, the merged ontology based on this alignment might contain unsatisfiable classes. The number of unsatisfiable classes induced by an alignment can be used as a global alignment evaluation metric. Let U be the set of unsatisfiable classes in $\mathcal{O}_1 \cup \mathcal{O}_2$. Let U_A be the set of unsatisfiable classes in $\mathcal{O}_A$. Let $\natural \mathcal{O}_A$ be the number of classes in $\mathcal{O}_A$, *i.e.* the number of classes in $\mathcal{O}_1 \cup \mathcal{O}_2$. The *alignment coherency evaluation metric* is defined as

$$H_{\text{coherence}}(A) = \begin{cases} \frac{|U_A| - |U|}{\natural \mathcal{O}_A - |U|} & \text{if } |U| \neq \natural \mathcal{O}_A \\ 1 & \text{otherwise} \end{cases} \tag{4.46}$$

This metric does only account for the number of classes that become unsatisfiable because of the alignment, and disregards the classes that were originally unsatisfiable in the two ontologies. In the case where all classes were originally unsatisfiable, the measure computes to 1, *i.e.* highest similarity, since the alignment has no impact on coherency in this case.

Structural Preservation

Motivated by the observation that two ontologies modelling the same domain are typically structured similarly with respect to their taxonomic (subsumption) and meronomic (part-of) hierarchies, Joslyn *et al.* propose a metric for evaluating ontology alignments based on structural preservation [71]. Their proposed metric is based on the notion of *distance* between entities according to those hierarchies. Informally, for two ontologies in the same domain, any two entities from the first ontology that have a close distance should correspond to two entities from the second ontology that have a close distance as well. The same holds for entities, which are farther apart.

The authors formalise their metric using concepts from order theory and the property of taxonomic / meronomic hierarchies in ontologies being partially ordered sets. For each pair of entities in an ontology the "lower cardinality-based distance" is computed[12], which is a metric based on the entities' successor sets and the intersection of those sets. Each entity pair's lower cardinality-based distance is subsequently normalised with respect to the size of the ontology. Based on these distances known for all entity pairs in each ontology, a *discrepancy* can be computed for any pair of correspondences in an alignment, and consequently for the complete alignment by averaging over all pairs of correspondences.

The authors evaluate their similarity metric using the *anatomy* ontologies from the Ontology Alignment Evaluation Initiative (OAEI) 2008 (cf. Section 7.2), and identified a positive correlation of F-measure scores of the participating systems and the alignment discrepancy.

For any two correspondences $C = \langle e_1, e_2 \rangle$ and $D = \langle f_1, f_2 \rangle$ contained in an alignment A, let

$$\delta(C, D) = |\bar{d}(e_1, f_1) - \bar{d}(e_2, f_2)| \qquad (4.47)$$

be the *correspondence discrepancy*, where $\bar{d}(e_i, f_i)$ ($i \in \{1, 2\}$) is the normalised lower distance between the entities e_i and f_i within ontology $\mathcal{O}_i$. Joslyn *et al.* suggest to normalise the distance to the size of the ontology in order to obtain a *relative* distance[13]. Following their definition, the nor-

---

[12]The reason for preferring the *lower* cardinality-based distance to the *upper* one is based on the assumption that ontologies are typically "down-branching", meaning that entities tend to have more subentities than superentities.

[13]Since the authors disregard that fact that different entity types can form hierarchies, and thus structural preservation needs to be considered with respect to those entity types, the authors' original formalisation has been adapted in this respect.

malised lower distance for e, $f \in \text{voc}_t(\mathcal{O})$ is $\bar{d}(e, f) = \frac{d(e, f)}{\sharp_t \mathcal{O} - 1}$ for any two entities of the same type. Informally, the lower cardinality-based distance $d(e, f)$ computes as the difference between the total number of subentities of e and f (including duplicates), and the largest number of subentities *shared* by both e and f. This denotes a distance metric on e and f, which is lower if the entities share a larger fraction of their subentities with one another, and thus are closer to each other in the subsumption hierarchy. (See the original work of Joslyn *et al.* [71] for details.)

Eventually, the *structural preservation evaluation metric* is defined using the average correspondence discrepancy of all pairs of correspondences in an alignment A:

$$H_{\text{structPreserv}}(A) = 1 - \frac{\sum_{C, D \in A} \delta(C, D)}{\binom{|A|}{2}} \tag{4.48}$$

Criss-cross Alignment

An evaluation metric analogous to the criss-cross correspondence similarity can be applied on the alignment level. The notion of "crossing" correspondences is used as in Equation (4.41). The portion of correspondences that cross other correspondences can be used as a global alignment evaluation metric.

Let $\mathcal{O}_1$ and $\mathcal{O}_2$ be two ontologies, and let A be an alignment between them. The set of correspondences that cross other correspondences in the alignment is defined as

$$\begin{aligned} X \quad = \quad & \{\langle e_1, e_2 \rangle \in A \mid \exists \langle f_1, f_2 \rangle \in A \text{ such that} \qquad (4.49) \\ & f_1 \in \text{sub}_{\mathcal{O}_1}(e_1) \wedge f_2 \in \text{sup}_{\mathcal{O}_2}(e_2) \text{ or} \\ & f_1 \in \text{sup}_{\mathcal{O}_1}(e_1) \wedge f_2 \in \text{sub}_{\mathcal{O}_2}(e_2) \} \end{aligned}$$

The *criss-cross alignment evaluation metric* is defined as

$$H_{\text{crissCross}}(A) = 1 - \frac{|X|}{|A|} \tag{4.50}$$

4.2. Similarity Aggregation

In order to compute a single evaluation value for a correspondence or an alignment, the individual evaluation scores discussed in Section 4.1 need to be aggregated.

Definition 4.10. An *aggregation function* is a function $\Gamma : \mathbb{R}^n \times \mathbb{R}^n \to \mathbb{R}$ computing an aggregation $\Gamma(\vec{f}, \vec{\omega})$, where $\vec{f}$ is a vector of evaluation scores, and $\vec{\omega}$ is a vector of weights.

4.2.1. Maximum Aggregation

A simple aggregation function is a projection of the highest evaluation score. There is no interpretation of weights for this aggregation function. The *maximum aggregation* is defined as

$$\Gamma_{\max}(\vec{f}, \vec{\omega}) = \max\{f_1, \ldots, f_n\} \tag{4.51}$$

This aggregation method is greedy in the sense that it disregards all individual evaluation scores apart from the best one. There are few use cases, where several *alternative* evaluation metrics are to be aggregated, and only the highest score should be considered, *e.g.* the evaluation of different pairs of annotation values. In case the evaluation metrics reflect complementary evaluation criteria, it is not reasonable to use this method.

4.2.2. Weighted Average Aggregation

Another straightforward combination of evaluation scores is the computation of a weighted mean. The *weighted average aggregation* is defined as

$$\Gamma_{\mathrm{weightAvg}}(\vec{f}, \vec{\omega}) = \frac{\sum_{i=1}^{n} \omega_i f_i}{\sum_{i=1}^{n} \omega_i} \tag{4.52}$$

This aggregation method allows full flexibility for assigning the relative importance of the evaluation scores to be aggregated. However, this flexibility comes at the cost of a significant configuration overhead when determining the weights, and thus predicting the relative importance of evaluation metrics. In particular this becomes problematic when the same weight configuration is used for aligning different ontologies. Since ontologies can have different characteristics, different evaluation metrics are important, and the weight configuration might not be transferable.

4.2.3. Ordered Weighted Average Aggregation

A self-adaptive solution for weighted average aggregation is to re-order weights according to the evaluation scores. The *ordered weighted average aggregation* has been described and successfully applied to ontology alignment by Ji *et al.* [70]. For $\vec{f} = (f_1, \ldots, f_n)$, let $\rho : \{1, \ldots, n\} \rightarrow \{1, \ldots, n\}$ be a permutation, such that for all $1 \leq i < j \leq n$ holds $f_{\rho(i)} \leq f_{\rho(j)}$. The *ordered weighed average aggregation* is defined as

$$\Gamma_{\mathrm{owa}}(\vec{f}, \vec{\omega}) = \frac{\sum_{i=1}^{n} \omega_i f_{\rho(i)}}{\sum_{i=1}^{n} \omega_i} \tag{4.53}$$

This aggregation method does not allow for assigning a fixed weight to a particular evaluation metric, but to assign a fixed weight to each position in the reordering of evaluation scores. Prior to the computation of this aggregation, the evaluation scores are arranged in descending order. Note that this does not necessarily specify that the highest evaluation score gets the highest weighting, since the weight order is fixed.

5. Ontology Alignment using Biologically-Inspired Optimisation Techniques

This chapter presents the core contribution of this work. The main results are published in *Information Sciences* Volume 192, June 2012 [16], as well as at the *7th International Symposium on Foundations of Information and Knowledge Systems*, March 2012 [19].

Two applications of biologically-inspired optimisation techniques for the ontology alignment problem are presented:

- An *Evolutionary Algorithm*

- A *Discrete Particle Swarm Optimisation* algorithm

Regarding the Evolutionary Algorithm, instead of implementing a particular variant directly, a hybrid approach incorporating ideas from Evolutionary Programming [52] and population-based Extremal Optimisation [105] was developed. The reason is that for ontology alignment the influence of solution components, *i.e.* correspondences, is expected to be relatively significant for the overall solution quality. Extremal Optimisation works mainly on the solution component level, while Evolutionary Programming considers solutions more globally. Other hybrids, *e.g.* EP-SOC [83] have shown to be successful in various application domains. However, the ontology alignment problem has several peculiarities, such as validity constraints that motivated the development of a special purpose Evolutionary Algorithm. Similar adjustments according to special purpose applications of Evolutionary Algorithms have been reported previously [29, 126, 83, 106].

Particle Swarm Optimisation [76, 116] as a second approach for tackling the ontology alignment problem was motivated by the fact that the social component in the Swarm Intelligence paradigm is considered a major advance in the area of population-based optimisation [75]. A structurally similar problem to ontology alignment, which has successfully been addressed using a Discrete Particle Swarm Optimisation approach [32, 33] was used as a basis for the presented algorithm.

The general procedure of applying (iterative) biologically-inspired optimisation algorithms to a particular problem is as follows:

1. The problem must be represented as an *optimisation problem*

2. An *objective function* must be developed in order to assess candidate solutions

3. A suitable *solution representation* format must be chosen

4. Suitable *update operations* must be defined

5. A *termination criterion* must be specified

As *termination criterion*, the presented algorithms use a specified maximum number of iterations that have to be completed, which is set sufficiently high in order to allow the algorithms to converge. In fact, other termination criteria, such as elapsed wall-clock time, a solution quality threshold, or stagnation of solution improvement could be used to determine or improve the runtime of the algorithm.

In the previous Section 2.2.2 the ontology alignment problem was formally introduced as an *optimisation problem* with optimality defined according to an alignment quality metric. In the following Section 5.1 this quality metric is turned into an *objective function*. Section 5.2 introduces two solution representation formats, namely the correspondence set representation and the correspondence permutation representation that are used in the presented algorithms. The Evolutionary Algorithm and the Discrete Particle Swarm Optimisation algorithm used for solving the ontology alignment problem are presented in Section 5.3. Section 5.4 concludes the chapter with a discussion about the disregard of recombination operators, as well as a comparison of the two presented approaches.

5.1. Objective Function

A solution to the ontology alignment problem is an alignment (cf. Section 2.2.2). Independent of the way a metaheuristic represents a solution internally, there must be an objective function used to assess the quality of the solution. The approach presented in this work tackles the ontology alignment problem as a *single-objective* optimisation problem, *i.e.* a single function evaluates an alignment and hence allows for comparing any two alignments between the same ontologies according to this function.

The alignment quality function from Definition 2.9 serves as objective function. Complying with the flexibility regarding various ontology characteristics to be considered (cf. Conjecture 2 in Section 1.1), the objective function is instantiated using similarity metrics and aggregation functions from the "toolbox" presented in Chapter 4. To this end, there are evaluation metrics on the alignment level, as well as evaluation metrics on the correspondence level.

Typically the evaluation scores for the single correspondences (both local and contextual) have a significant contribution to the overall alignment evaluation score, and thus to the alignment quality. An objective function that is set up accordingly, thus implicitly computes evaluation scores for the single correspondences. These correspondence evaluation scores can be used to assign for each correspondence $C \in A$ its confidence value $\iota(C)$.

The assumption that for each correspondence of an alignment $C \in A$ its confidence $\iota(C)$ contributes positively to the alignment quality $F(A)$, makes ι a valuable *support heuristic* to be used by the presented metaheuristic applications. Speaking in terms of metaheuristics, the *fitness* of an individual in the population is equal to its objective function.

5.2. Solution Representation

The way candidate solutions are represented is an important factor for the efficient execution of the algorithm. On the one hand, the solution representation should facilitate the evaluation of a solution via the objective function. On the other hand, it should facilitate the maintenance of solution validity constraints[1] throughout the optimisation run (cf. Definition 2.7). The section introduces two solution representation formats, which are later used in the presented alignment algorithms, namely the *correspondence set* and the *correspondence permutation*.

---

[1]The presented approaches do not allow infeasible solutions during the course of the optimisation run.

5.2.1. Correspondence Set Representation

The *correspondence set* representation is the straightforward way of representing a solution to the alignment problem, namely as a set of correspondences. Hence the correspondence set representation coincides with the definition of a valid alignment given in Definition 2.7.

An advantage of this representation is that it allows for easy assessment of the represented alignment with respect to the objective function. A disadvantage is that it requires the update operation U to be aware of global alignment constraints, such that only valid candidate alignments are produced by U. In other words, U must be aware of the validity criteria given in Definition 2.7, *i.e.* U must be specified in a way that only valid "?:?" alignments are generated when applying the update operation.

Example. Given the two ontologies from Figure 1.1, the correspondence set representation of a solution to the ontology alignment problem is

$\{\langle$Thing, Thing$\rangle, \langle$Entry, Publication$\rangle, \langle$Article, Article$\rangle,$
$\langle$Book, Book$\rangle, \langle$Inproceedings, InProceedings$\rangle,$
$\langle$Manual, Manual$\rangle, \langle$Misc, Misc$\rangle, \langle$Phdthesis, Thesis$\rangle,$
$\langle$Proceedings, Proceedings$\rangle, \langle$TechReport, TechnicalReport$\rangle,$
$\langle$Unpublished, Unpublished$\rangle\}$

5.2.2. Correspondence Permutation Representation

The *correspondence permutation* representation denotes a novel data structure that does naturally exclude invalid solutions (alignments).

Definition 5.1. For two ontologies $\mathcal{O}_1$ and $\mathcal{O}_2$ and an entity type $t \in \mathcal{T}$, let m be the number of entities of type t in $\mathcal{O}_1$, and let n be the number of entities of type t in $\mathcal{O}_2$. A *correspondence permutation* is a function $\pi_t : \{1, \ldots, m\} \rightarrow \{1, \ldots, n\} \cup \square$, such that for all $1 \leq i < j \leq m$ with $\pi_t(i), \pi_t(j) \neq \square$ holds $\pi_t(i) \neq \pi_t(j)$. (For the sake of brevity the type specifier t can be omitted if the type is irrelevant for a particular argument.)

A correspondence permutation can be understood as an array of m elements: $\pi(1)\ \pi(2)\ \cdots\ \pi(m)$. Thereby, each of the numbers $1 \ldots n$ can occur at most once in this array, whereas $\square$ can occur arbitrarily often.

Assuming the array index represents the entity index of (the smaller) ontology $\mathcal{O}_1$ and the array element at position j denotes the index $\pi(j)$ of

1	2	3	4	5	6	7	8	9	10	11	12	13	14	15	16
1	15	16	17	□	□	□	□	18	19	□	20	25	21	23	29

Figure 5.1.: Example of a correspondence permutation using indexes from Figure 1.1. The array index represents the entity index of the ontology from Figure 1.1a, while the array elements represent entity indexes of the ontology from Figure 1.1b.

the entity in ontology $\mathcal{O}_2$, it is straightforward that each array entry represents a correspondence $\langle e_j, f_{\pi(j)} \rangle$ if $f_{\pi(j)} \neq \square$, and no correspondence for e_j otherwise.

Example. Figure 5.1 shows an example of a correspondence permutation that represents one possible alignment for the two ontologies from Figure 1.1. (The example represents the same solution as the example for the correspondence set representation in Section 5.2.1.)

Since by Definition 2.5 correspondences can only exist between entities of the same type, a solution representation would require $|\mathcal{T}|$ correspondence permutations, one for each entity type.

If for any $t \in \mathcal{T}$, $\sharp_t \mathcal{O}_1 \neq \sharp_t \mathcal{O}_2$, the index sets for $\mathrm{voc}_t(\mathcal{O}_1)$ and $\mathrm{voc}_t(\mathcal{O}_2)$ should be swapped, such that $m < n$ for increasing memory efficiency. (If the array index would reflect the entity index of the larger type-restricted vocabulary, such that $m > n$, the correspondence permutation would needlessly contain at least $m - n$ "$\square$" values.) For the sake of simplicity in the remainder of this chapter this optimisation is ignored.

5.3. Iterative Convergence

Biologically-inspired optimisation algorithms work iteratively. Thereby the desired behaviour is that they converge towards a (near-)optimal solution in a guided fashion, *i.e.* convergence happens significantly faster than for randomly sampling the solution space. On the other hand, the algorithm should be robust against getting stuck in local optima, resulting in a premature stagnation of the convergence.

In the presented Evolutionary Algorithm, the convergence is controlled by mutation and selection operators that update the individuals of the population in each iteration. For the Discrete Particle Swarm Optimisation algorithm, movement operations are responsible for convergence by placing the particles at new promising positions in the solution space.

5.3.1. Mutation and Selection in an Evolutionary Algorithm

This section presents a novel Evolutionary Algorithm for solving the ontology alignment problem. The algorithm is a hybrid approach taking advantage of features of Evolutionary Programming and population-based Extremal Optimisation [105, 83]. The following paragraphs pick up the generic notations for population-based optimisation techniques as introduced in Section 2.3 and focus on the mutation and selection operators that are the decisive factor in this approach.

Formal Definitions

As presented generally for Evolutionary Algorithms in Section 2.3, a population $\langle I, p \rangle$ of species (candidate solutions) is exposed to an environment (problem space). In a number of n iterations (generations) the population continuously changes, such that better adapted individuals survive and reproduce themselves, while less adapted individuals become extinct. Two crucial operations are involved in this approach, namely *mutation* and *selection*. While mutation is applied to single individuals, selection is applied to the entire population.

Each species represents a *candidate alignment* between two ontologies $\mathcal{O}_1$ and $\mathcal{O}_2$ using the *correspondence permutation* representation introduced in Section 5.2.2. Thereby, for each entity type $t \in \mathcal{T}$ there is a correspondence permutation $\pi_t : \{1, \ldots, m_t\} \to \{1, \ldots, n_t\}$ where $m_t = \sharp_t \mathcal{O}_1$ and $n_t = \sharp_t \mathcal{O}_2$. For every species $x \in I$ the position in the problem space is defined as

$$p(x) = \bigcup_{t \in \mathcal{T}} \left\{ \langle e_j, f_{\pi_t^{(x)}(j)} \rangle \mid j \in \{1, \ldots, m_t\}, \pi_t^{(x)}(j) \neq \square \right\} \qquad (5.1)$$

Hereby, $\pi_t^{(x)}$ denotes the correspondence permutation for type t represented by species x. The set $p(x)$ of correspondences represented by the correspondence permutations for the different entity types in $\mathcal{O}_1$ and $\mathcal{O}_2$ is called the configuration of the species. Note that this configuration is also an alignment. Without loss of generality, the remainder of this section considers only a single entity type for the sake of brevity.

Mutation Operators. In traditional Evolutionary Algorithms as those presented in Section 2.3 mutation of species is a purely random modification of a solution component[2] in order to explore previously unexplored regions of the problem space. This approach is valid and expedient for black box optimisation problems, where there is no suitable support heuristic in order to evaluate solution components. However, in many application domains, there is some knowledge about the structure of the problem, and thus about the contribution of solution components to the overall solution. In these cases, it is common practise to use adapted (informed) mutation operators in order to accelerate the convergence. For instance in an application of Evolutionary Programming for the problem of multi-sequence DNA/RNA alignment, Chellapilla and Fogel [29] implement a mutation operator that uses an already good part of the solution and tries to improve its neighbourhood. In a similar fashion for the problem of ontology alignment, the confidence of single correspondences in an alignment can be used as a support heuristic that has an impact on the mutation. (See discussion in Section 5.1.)

In the following paragraphs, two mutation operators are introduced that are adapted to ontology alignment and exploit the correspondence confidence values as a support heuristic: the *swap* operator and the *exchange* operator.

Swap Operator. The *swap* operator u_s turns a correspondence permutation π into a new one π' by picking two indexes j and k, such that $1 \leq j < k \leq m$ and $\pi(j), \pi(k) \neq \square$. Subsequently, it transforms the correspondence permutation as follows by setting $\pi'(j) = \pi(k)$ and $\pi'(k) = \pi(j)$ as well as $\pi'(l) = \pi(l)$ for all $l \notin \{j, k\}$.

Let j be a candidate index to be selected for being swapped. If $\pi(j) \neq \square$, let $C_j = \langle e_j, f_{\pi(j)} \rangle$ be the correspondence represented by j and let $\iota(C_j)$ be its confidence. Further let $\mu = \frac{1}{|A|} \sum_{D \in A} \iota(D)$ the mean confidence of all correspondences in the alignment. Index j is selected for being swapped with probability

$$p_{swap}(j) = p_{swapSelect}(\iota(C_j), \mu) \cdot \varrho \qquad (5.2)$$

where ϱ is a parameter used to limit the number of swap operations hap-

---

[2]Typically solutions are represented as bit strings, where the mutation operator randomly flips bits.

pening[3] and $p_{swapSelect}(x, \mu)$ is defined as follows:

$$p_{swapSelect}(x, \mu) = \begin{cases} 1 - (\text{sigmoid}_\mu(x) \cdot 2\mu) & \text{if } x \leq \mu \\ 1 - (\text{sigmoid}_\mu(x) \cdot 2(1 - \mu) + 2\mu - 1) & \text{otherwise} \end{cases} \tag{5.3}$$

where

$$\text{sigmoid}_\mu(x) = \frac{1}{1 + e^{-10(x-\mu)}} \tag{5.4}$$

is the sigmoid function shifted, such that its inflection point is at μ. Figure 5.2 illustrates how the $p_{swapSelect}$ depends on the particular correspondence's confidence and the mean confidence of all correspondences.

The rational behind this sort of probability computation is that depending on the evaluation metrics used to determine the confidence of a correspondence, the confidence values are not distributed equally in the range $[0, 1]$. In case no suitable evaluation metrics are available, or the used evaluation metrics do not perfectly address the required ontology characteristics, the evaluation scores would centre around a comparatively low value. The computation of the $p_{swapSelect}$ probability takes the mean confidence into account and favours correspondences of a confidence below average to be swapped, and reduces the probability for correspondences with a confidence above average.

Note that the swap operator does not change the size of the alignment, since no correspondences are removed from or added to the alignment. Moreover, the entities from both ontologies participating in the alignment, will be the same before and after executing the swap operator.

Exchange Operator.   The *exchange* operator u_e also transforms a correspondence permutation π into a new one π'. The operator modifies the configuration of a species by removing an existing correspondence (setting its correspondence permutation value to □), by adding a new correspondence (setting a □ correspondence permutation value to the index of an entity that is not already participating in another correspondence), or by replacing an existing entity with another entity that is not already participating in another correspondence.

---

[3]The parameter ϱ should typically be a small value (good results were achieved with values between 0.1 and 0.3), since it limits impact of the swap operator and makes this mutation less "aggressive".

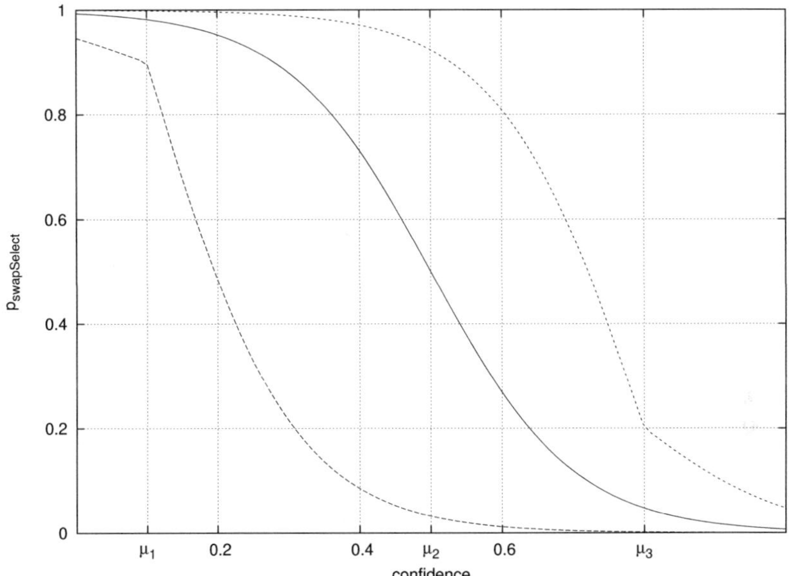

Figure 5.2.: Probability to select an index as candidate for swapping. The probability is computed according to Equation (5.3) and depends on the confidence of the correspondence represented by the index, and the mean confidence of all represented correspondences. The plots show the probabilities for three different mean confidences ($\mu_1 = 0.1$, $\mu_2 = 0.5$, $\mu_3 = 0.8$).

Let $R = \{1, \ldots, n\} \setminus \{\pi(1), \ldots, \pi(m)\}$ be an archive of entity indexes of $\mathcal{O}_2$ not currently participating in a correspondence. The exchange operator modifies the correspondence permutation C according to the decision tree shown in Figure 5.3. The action to perform is selected for each correspondence permutation index j independently.

The top-level branching is done depending on whether the current correspondence permutation index actually represents a correspondence. If the index does not represent a correspondence, *i.e.* $\pi(j) = \square$, a decision is taken whether to add a correspondence to be represented by this index. This modification is performed with probability p_{setV}. If the index j actually represents a correspondence, *i.e.* $\pi(j) \neq \square$, a first decision is taken with probability p_{change} whether to perform a change at all. In the pos-

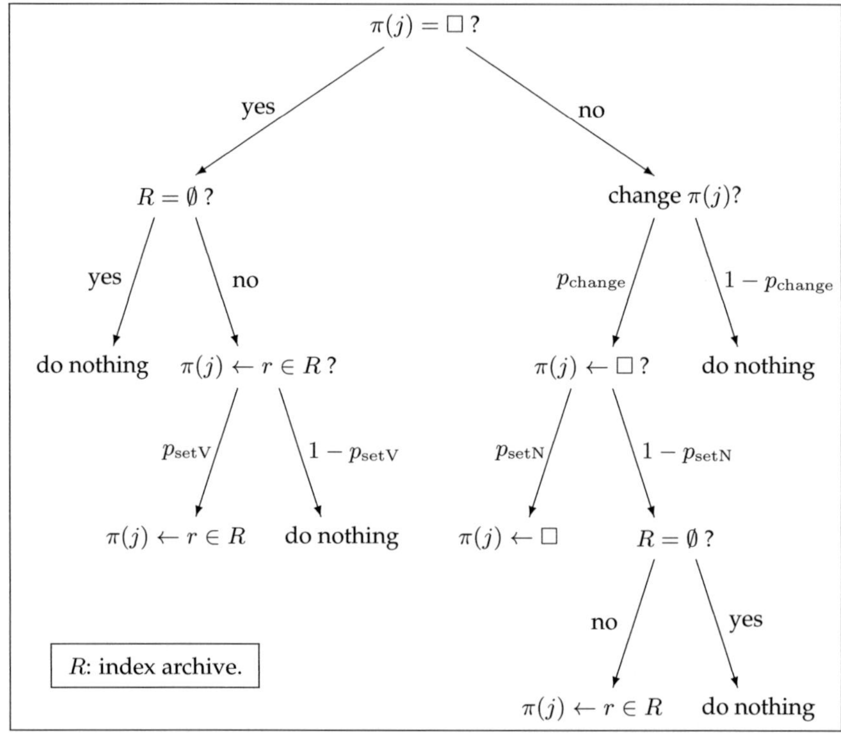

Figure 5.3.: Decision tree for executing the exchange operator.

itive case, probability p_{setN} determines whether the correspondence will be removed or whether the entity $f_{\pi(j)} \in \mathcal{O}_2$ corresponding to $e_j \in \mathcal{O}_1$ will be replaced by a random entity from $\mathcal{O}_2$ that is not currently participating in any correspondence.

The probabilities p_{change}, p_{setN}, and p_{setV} depend on the number of iterations that have elapsed. Let i be the current iteration, and let i_{max} be the maximum number of iterations to perform.

The probability p_{setV} to add a new correspondence for an entity $e_j \in \mathcal{O}_1$ currently not participating in any correspondence is defined as

$$p_{setV}(i) = -\frac{i}{i_{max}}(b - a) + b \tag{5.5}$$

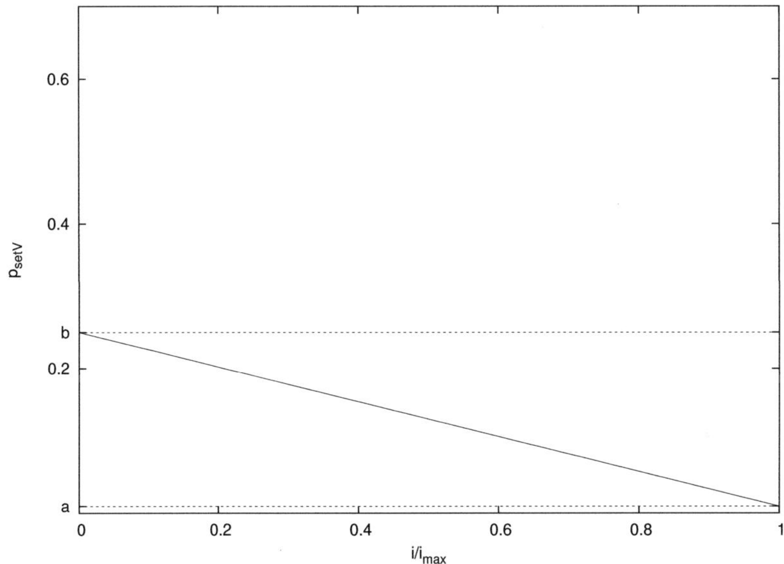

Figure 5.4.: Exchange operator: probability p_{setV} to add a correspondence.

which is a decreasing linear function depending on the iteration progress i/i_{max}. It has two constant parameters a and b, determining the lower and upper bound of the probability, respectively. Figure 5.4 shows a plot of p_{setV} depending on the iteration progress.

The rational behind the probability decreasing in the course of the iterations is that the added correspondence is randomly chosen. There is no guarantee about the quality of the newly generated correspondence. Towards the end of the optimisation run with few iterations left, there are few chances that a chosen correspondence of low quality will be corrected by future mutations. So decreasing the probability throughout the iterations makes the algorithm more conservative towards the end.

The lower and upper bound parameters a and b should be set according to the intuition of the probability decrement. In particular, the lower bound a should be close to 0. The upper bound b can incorporate knowledge about the expected overlap of the ontologies, since the higher its value, the more likely "□" entries in the correspondence permutation will be replaced by entity indexes. This leads to a more complete overlap rep-

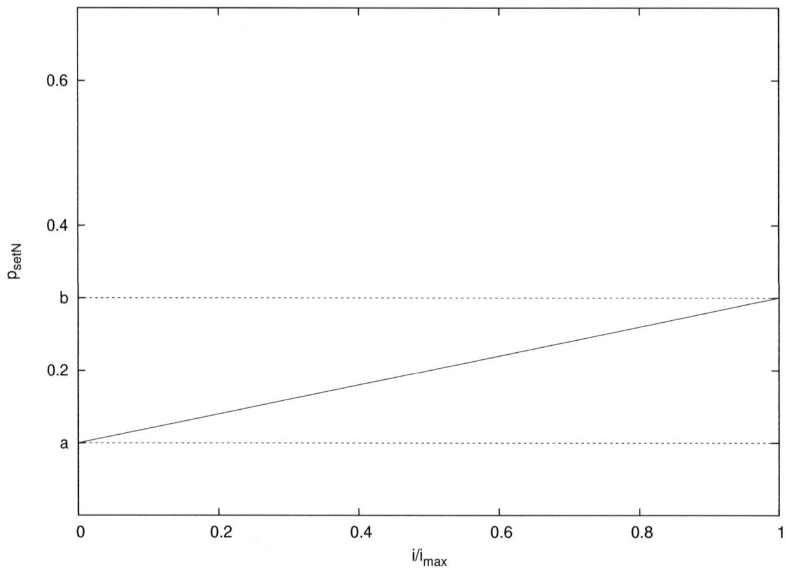

Figure 5.5.: Exchange operator: probability p_{setV} to remove a correspondence.

resented by the alignment. Empirical studies have shown that values of b between 0.2 and 0.5 lead to good results.

The probability p_{setN} to remove an existing correspondence for an entity $e_i \in \mathcal{O}_1$ is defined as

$$p_{setN}(i) = \frac{i}{i_{max}}(b - a) + a \tag{5.6}$$

which is an increasing linear function depending on the iteration progress i/i_{max}. It has two constant parameters a and b, determining the lower and upper bound of the probability, respectively. Figure 5.5 shows a plot of p_{setN} depending on the iteration progress.

The rational behind the probability increasing in the course of the iterations is that towards the end of the optimisation run with few iterations left, there are few chances that a bad correspondence will be improved by future mutations. So increasing the probability to remove bad correspondences throughout the iterations cleans out the alignment from bad correspondences thus decreasing its size. This behaviour is particularly desired for ontologies with only a partial overlap.

Following this intuition, the lower and upper bound parameters a and b should be set accordingly. Removing correspondences should be a rare operation at the beginning of the optimisation run, so a should be set to a value close to 0. The upper bound parameter b determining the probability at the end of the optimisation run can incorporate knowledge about the expected overlap of the ontologies, since a higher value of b causes the alignment to shrink towards the end of the algorithm execution[4]. Empirical studies have shown that values of b of between 0.1 and 0.5 lead to good results.

The probability p_{change} to change an existing correspondence for an entity $e_j \in \mathcal{O}_1$ depends not only on the iteration i, but also on the confidence $\iota(\langle e_j, f_{\pi(j)} \rangle)$ of this correspondence. It is defined as

$$p_{change}(i, \iota) = \begin{cases} 1 - \frac{\iota}{g_a(i)} & \text{if } g_a(i) > \iota \\ 0 & \text{otherwise} \end{cases} \tag{5.7}$$

where

$$g_a(i) = (1 - a)\frac{i}{i_{max}} + a \tag{5.8}$$

Similar to the p_{swap} probability the p_{change} probability depends on the confidence of the correspondence considered for a change as a support heuristic. The probability for an existing correspondence to be changed decreases with a higher confidence. As a second factor, the iteration progress influences the probability. At the beginning of the optimisation run, correspondences with a confidence above a threshold a are never changed. In the course of the iterations this threshold increases towards 1 in the last iteration. Figure 5.6 shows a plot of p_{change}.

The rational behind the definition of this probability function is as follows. On the one hand, correspondences with low confidence should always have a higher probability to be changed than those with a high confidence. On the other hand, the probability to change a correspondence should increase with the iteration progress also for better correspondences, for the following two reasons:

- After the decision to perform a change it is decided whether to remove the correspondence or replace its second entity with proba-

---

[4]Since the decision of whether to remove a correspondence also depends on its confidence (cf. Equation (5.7)), the expected confidence value above which correspondences are regarded "good" also influences the choice of the value of parameter b.

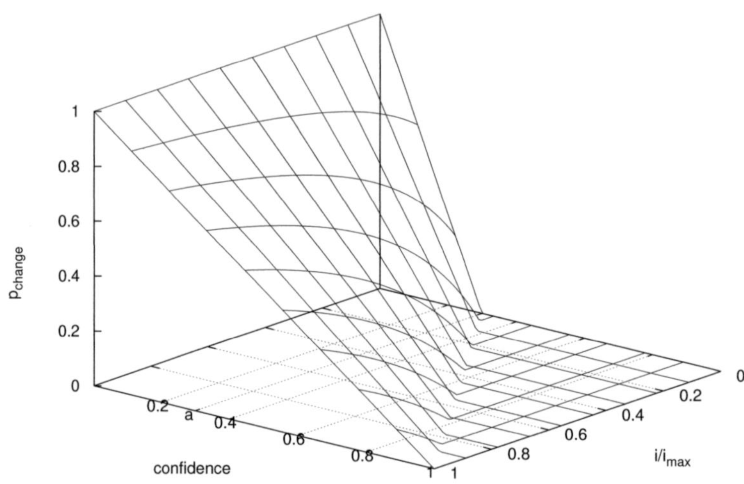

Figure 5.6.: Exchange operator: probability p_{change} to change an existing correspondence. a denotes the initial confidence threshold.

bility p_{setN}. This latter probability is increasing, such that in combination with p_{change} changes will more likely remove than modify
correspondences. This will sort out correspondences of low and
medium confidence and help improving the average confidence of
correspondences in the alignment.

- In early iterations correspondences of medium confidence will be
  kept, since there will be chances in later iterations that these correspondences will be altered by the swap operator. Conversely, it
  is more likely that towards the end of the optimisation run correspondences of low and medium confidence have undergone several swap operations already. Correspondences that remain with
  low or medium confidences then have the chance to be replaced or
  removed.

The dependence on the elapsed number of iterations has previously
been studied by Thomsen [126] in terms of so-called *annealing* schemes.

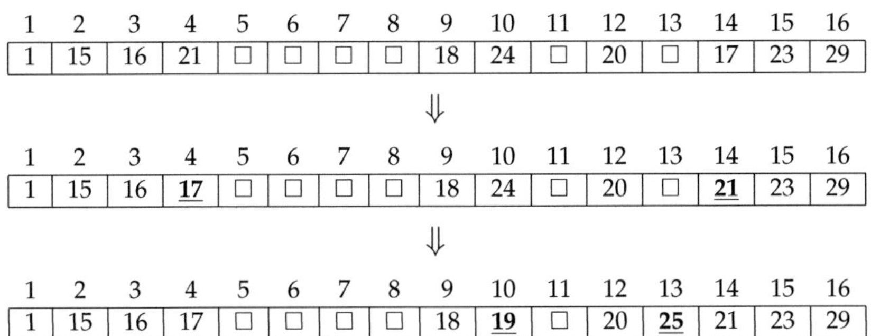

Figure 5.7.: Example application of the *swap* operator followed by the *exchange* operator on a correspondence permutation.

The underlying idea is that with the iteration progress the found solution becomes more stable, and appropriate actions can be taken, such as limiting the amplitude of changes done by the mutation operator, or, as in this case, increasing the chance to clean out bad solution components towards the end of the execution.

Example. In order to illustrate the effects of the two mutation operators presented in the previous paragraphs, a correspondence permutation is mutated to achieve the one from Figure 5.1 using the two operators. Figure 5.7 illustrates the application of the *swap* operator, followed by an application of the *exchange* operator. In the first step indexes 4 and 14 are selected for being swapped. In the second step, index 10 is selected for being exchanged, and since its value is not □, a decision is taken whether to replace the previous value, 24, with □ or another random value from the archive. In this case the decision is to replace it with another value, 19. For the array index 13 the exchange operator is also applied, and since its value is □ and the decision to set a new value was positive, it is replaced by a random value from the archive, here 25.

Selection. Apart from mutation the second important operation in evolutionary processes modelled by this evolutionary algorithm is selection. In this algorithm a simplified linear rank-based selection [58] is applied, which is controlled by a selection ratio parameter ζ. For a population $\langle I, p \rangle$ and an objective function F (cf. Section 5.1 and Definition 2.9) a ranking is defined as a bijective function $r : I \rightarrow \{1, \ldots, |I|\}$, such that for any two x_i, $x_j \in I$ with $1 \leq i < j \leq |I|$ holds $r(x_i) < r(x_j)$ iff $F(p(x_i)) < F(p(x_j))$. Let $E = \{x \in I \mid r(x) \leq \zeta \cdot |I|\}$ be the set of species to become extinct. Let $S = \{x \in I \mid r(x) > (|I| - \zeta \cdot |I|)\}$ be the set of species to be allowed to reproduce. Each species $x \in S$ reproduces by creating exactly one offspring y, which is a clone of x, such that $p(y) = p(x)$. Note that the population size remains constant due to $|E| = |S|$. Let s be a selection function transforming a population $\langle I, p \rangle$ into a new one $\langle I', p' \rangle$, such that $I' = (I \setminus E) \cup \{x \mid x$ is a clone of $y \in S\}$ and for all $x \in I'$

$$p'(x) = \begin{cases} p(x) & \text{if } x \in I \setminus E \\ p(y) & \text{if } x \text{ is an offspring of } y \in S \end{cases} \qquad (5.9)$$

The first case represents the preservation of the assignment of a "surviving" species x, that is if $x \in I$ and $x \in I'$. The second case represents the assignment of the "new" species, *i.e.* those that were created by reproduction. They are assigned the same position in the solution space as their parent $y \in S$.

The presented Evolutionary Algorithm for ontology alignment conducts several atomic mutation operations before carrying out a selection step. This reflects an intensified search by the different species in different areas of the problem space, before a poorly performing species becomes extinct. Reducing the impact of selection like this is motivated by the following observation. Experience has shown that for this application of Evolutionary Computation the mutation operation has a stronger emphasis on the quality of the obtained solutions. The reason is that the influence of the support heuristic of correspondence level evaluation scores is relatively significant for the overall alignment evaluation. The mutation operators are influenced by those evaluation scores for correspondences and modify the represented alignment on the correspondence level. Reducing the frequency of selection operations has also been shown to be successful by Randall [105] for a population-based Extremal Optimisation algorithm. A similar selection strategy where the worst performing population members according to the objective function are replaced was introduced in the EPSOC algorithm by Lewis *et al.* [83].

Considering the periodic application of the selection function s, as well as the mutation operators u_s and u_e, the update of a population from iteration i to iteration $(i + 1)$ computes as

$$\langle I_{i+1}, p_{i+1} \rangle = U(\langle I_i, p_i \rangle) \qquad (5.10)$$

where $I_{i+1} = s(I_i)$ and $p_{i+1} = (u_e \circ u_s \circ s)(p_i)$.

Algorithm 5.1 Initialisation of Species

Require: $|I|$ the number of species
  for $i = 1$ **to** $|I|$ **do**
    for all $t \in \mathcal{T}$ **do**
      $m_t = \min\{\sharp_t \mathcal{O}_1, \sharp_t \mathcal{O}_2\}$
      $n_t = \max\{\sharp_t \mathcal{O}_1, \sharp_t \mathcal{O}_2\}$
      $R \Leftarrow \{1, \ldots, n_t\}$
      for $j = 1$ **to** m_t **do**
        $\pi_t(j) = \mathrm{rand}_U(R)$
        $R \Leftarrow R \setminus \pi_t(j)$
        Compute $\iota(\langle e_j, f_{\pi_t(j)}\rangle)$
      end for
    end for
  end for

Algorithm

This subsection presents an algorithm that computes an ontology alignment following the method presented in the previous paragraphs. The algorithm is split into three parts: an initialisation step, the evolution of the population, and an update procedure, *i.e.* the application of the two mutation operators for each species.

The computation of an alignment starts with an initialisation as presented in Algorithm 5.1. Hereby, each species is initialised with a random configuration of a correspondence permutation for each entity type.

The optimisation run is an iterative, guided evolution of the population as encoded in Algorithm 5.2. The total number of iterations is split into "sprints", where several consecutive mutations happen without a selection. This can be seen as a single mutation with intermediate confidence re-computations on the correspondence level. The rational behind this method, as well as the details about the selection step were given in the previous paragraphs.

The update of species according to the two mutation operators presented earlier is done according to Algorithm 5.3. Note that all species can be evaluated and updated in parallel.

Algorithm 5.2 Population Evolution

Require: $|I|$ the number of species
  i_{max} the number of iterations
  i_{sel} the number of selection steps
  ζ the selection ratio

  for $s = 1$ **to** i_{sel} **do**
    for $j = 1$ **to** $|I|$ **do**
      for $i = 1$ **to** $\lfloor i_{max}/i_{sel} \rfloor$ **do**
        Mutate species $x_j \in I$ according to Algorithm 5.3
        (using current iteration $i + s \cdot \lfloor i_{max}/i_{sel} \rfloor$)
      end for
      Compute $F(p(x_j))$
    end for
    Replace $\zeta \cdot |I|$ worst species according to F
    with $\zeta \cdot |I|$ best species according to F (cf. Equation (5.9))
  end for

5.3.2. Particle Movement in Swarm Optimisation

In this section a novel Discrete Particle Swarm Optimisation (DPSO) algorithm for solving the ontology alignment problem is introduced. The algorithm is motivated by a DPSO algorithm of Correa *et al.* [32] applied to the problem of attribute selection for a Naïve Bayes classifier (cf. Section 3.2.3). Correa *et al.* [32] show that the classical binary PSO has problems finding the optimal (smallest) number of attributes. In the case of ontology alignment there is the analogous problem of finding the largest number of correspondences in an alignment. For this reason, the approach presented here does not adopt the classical binary PSO, but instead uses a modified version of the DPSO by Correa *et al.* based on the correspondence set representation introduced in Section 5.2.1.

Formal Definitions

Recalling and extending the definitions from Section 2.3 a particle swarm is a population $\langle I, p \rangle$, which moves through the problem space in n iterations. In traditional PSO each particle updates its position in the problem space using a so-called *velocity* vector. This movement happens via a guided, randomised re-initialisation of each particle. Since this approach uses a modified *discrete* PSO, this idea is partially relaxed, and particles and velocities are defined following the approach of Correa *et al.* [32].

Each particle represents a *candidate alignment* between two ontologies $\mathcal{O}_1$ and $\mathcal{O}_2$ using the *correspondence set* representation introduced in Section 5.2.1. Particles can have different dimensionality, *i.e.* the number of correspondences in the alignment it currently represents, and hence differ from traditional PSO, where each particle has the same dimensionality. A dimensionality of zero means that a particle represents the empty alignment. For every particle $x \in I$ the position in the problem space is defined

---

Algorithm 5.3 Update of Species x_i (part 1)

Require: i the current iteration
1: $\forall t \in \mathcal{T}$:
2: π_t a correspondence permutation
3: $m_t = \min\{\sharp_t \mathcal{O}_1, \sharp_t \mathcal{O}_2\}$
4: $n_t = \max\{\sharp_t \mathcal{O}_1, \sharp_t \mathcal{O}_2\}$
5: $R_t \subseteq \{1, \ldots, n_t\}$ an archive of indexes of currently unused entities of type t from the bigger of $\mathrm{voc}_t(\mathcal{O}_1)$ and $\mathrm{voc}_t(\mathcal{O}_2)$.
6: **for all** $t \in \mathcal{T}$ **do**
7: $\triangleright$ swap operator:
8:       $S \Leftarrow \emptyset$
9:       **for** $j = 1$ **to** m_t **do**
10:           **if** $\mathrm{rand}_U < p_{swap}(j)$ **then**
11:              $S \Leftarrow S \cup \{j\}$
12:           **end if**
13:       **end for**
14:       $S' \Leftarrow S$
15:       $\pi'_t \Leftarrow \pi_t$
16:       **for all** $k \in S$ **do**
17:           $k' \Leftarrow \mathrm{rand}_U(S')$
18:           $\pi_t(k) \Leftarrow \pi'_t(k')$
19:           $S' \Leftarrow S' \setminus \{k'\}$
20:       **end for**
21:       **for all** $C \in \{\langle e_j, f_{\pi_t(j)} \rangle \mid j \in \{1, \ldots, m_t\}, \pi_t(j) \neq \square\}$ **do**
22:           Compute $\iota(C)$
23:       **end for**
24:       $\ldots$

Algorithm 5.4 Update of Species x_i (part 2)

25:        $\ldots$

26:   ▷ exchange operator (decision tree, cf. Figure 5.3):

27:        **for** $j = 1$ **to** m_t **do**

28:            **if** $\pi_t(j) = \square$ **then**

29:                **if** $\text{rand}_U < p_{setV}(i)$ **then**

30:                    $\pi_t(j) \Leftarrow \text{rand}_U(R_t)$

31:                    $R_t \Leftarrow R_t \setminus \{\pi_t(j)\}$

32:                **end if**

33:            **else**

34:                **if** $\text{rand}_U < p_{change}(i, \iota(\langle e_j, f_{\pi_t(j)}\rangle))$ **then**

35:                    **if** $\text{rand}_U < p_{setN}(i)$ **then**

36:                        $R_t \Leftarrow R_t \cup \pi_t(j)$

37:                        $\pi_t(j) \Leftarrow \square$

38:                    **else**

39:                        $\pi_t(j) \Leftarrow \text{rand}_U(R_t)$

40:                        $R_t \Leftarrow R_t \setminus \{\pi_t(j)\}$

41:                    **end if**

42:                **end if**

43:            **end if**

44:        **end for**

45:        **for all** $C \in \{\langle e_j, f_{\pi_t(j)}\rangle \mid j \in \{1, \ldots, m_t\}, \pi_t(j) \neq \square\}$ **do**

46:            Compute $\iota(C)$

47:        **end for**

48: **end for**

---

as a vector[5]

$$p(x) = \left\{ C_{(x,1)}, C_{(x,2)}, \ldots, C_{(x,k)} \right\} \tag{5.11}$$

where for each $j \in \{1, \ldots, k\}$, $C_{(x,j)}$ is a correspondence as defined in Definition 2.5. This set of correspondences is also called a *configuration* of the particle. Note that this configuration is also an alignment. Since the maximum number of correspondences in an alignment according to Equation (2.1) is

$$N = \sum_{t \in \mathcal{T}} \min\{\sharp_t\mathcal{O}_1, \sharp_t\mathcal{O}_2\}$$

---

[5]Note that the exact mathematical notation is violated and the vector is denoted with curly braces, as it can also be seen as a set.

the (variable) dimensionality of a particle is $k \in \{0, \ldots, N\}$. According to Definition 2.9, the fitness of a particle x is

$$F(p(x)) \tag{5.12}$$

Each particle x maintains the configuration of the best alignment it has ever represented with respect to F. This *personal best (pBest)* alignment of dimensionality $l \in \{0, \ldots, N\}$ is denoted by

$$p_{\text{personal}}(x) = \left\{ D_{(x,1)}, D_{(x,2)}, \ldots, D_{(x,l)} \right\} \tag{5.13}$$

where for each $j \in \{1, \ldots, l\}$, $D_{(x,j)}$ is a correspondence. Note that the number of correspondences can change during the iteration of the swarm (see later in this section). Hence the dimensionality l of the *pBest* configuration of a particle does not need to coincide with the dimensionality k of its current configuration. The *neighbourhood best (nBest)*, *i.e.* the best performing parameter configuration any particle in the neighbourhood of a particular particle x has ever represented with respect to F is denoted by

$$p_{\text{neighbour}}(x) = \left\{ D_{(x,1)}, D_{(x,2)}, \ldots, D_{(x,m)} \right\} \tag{5.14}$$

where for each $j \in \{1, \ldots, m\}$, $D_{(x,j)}$ is a correspondence. Its dimensionality is $m \in \{0, \ldots, N\}$.

To ensure a guided convergence towards an optimal alignment during the iterations, the influence of arbitrary random re-initialisation of each particle has to be restricted. To this end, the likelihood is raised that those correspondences in a particle are preserved, which *(i)* are evaluated best, and *(ii)* are also present in the personal (5.13) or neighbourhood (5.14) best alignment.

The *fitness vector* of a particle x is denoted by a 2-by-k array

$$\vec{F}_x = \begin{pmatrix} \iota_{(x,1)} & \iota_{(x,2)} & \cdots & \iota_{(x,k)} \\ C_{(x,j_1)} & C_{(x,j_2)} & \cdots & C_{(x,j_k)} \end{pmatrix} \tag{5.15}$$

associating a fitness $\iota_{(x,\mu)}$ to each correspondence $C_{(p,j_\mu)}$. The confidence ι as defined in Definition 2.6 is used to reflect the fitness of a correspondence. This is a support heuristic and does not replace the main objective[6] to optimise $F(p(x))$. The vector is ordered by its confidence values.

---

[6]The objective function $F(A)$ typically incorporates the confidence values $\iota(C), \forall C \in A$ as done by the correspondence computation alignment evaluator (cf. Equation (4.43)). This, however, is not necessarily required.

A *velocity vector* is defined as another 2-by-k array

$$\vec{V}_x = \begin{pmatrix} v_{(x,1)} & v_{(x,2)} & \cdots & v_{(x,k)} \\ C_{(x,l_1)} & C_{(x,l_2)} & \cdots & C_{(x,l_k)} \end{pmatrix} \tag{5.16}$$

mapping a proportional likelihood $v_{(x,\mu)}$ to each correspondence $C_{(x,l_\mu)}$. The vector is ordered by its proportional likelihoods. Proportional likelihoods are used to raise the probability of those correspondences to be preserved in a particle that are also present in the personal and neighbourhood best alignments. Initially, for each $C_{(x,l_\mu)}$, $v_{(x,\mu)}$ is set to 1. This initialisation is also done for new correspondences joining the particle during its movement. The update of the proportional likelihoods is then done in two steps, using two parameters $\beta \in \mathbb{R}^+$ and $\gamma \in \mathbb{R}^+$. Firstly, if $C_{(x,l_\mu)}$ is present in $p_{\text{personal}}(x)$, add β to $v_{(x,\mu)}$. If it is present in $p_{\text{neighbour}}(x)$, add γ to $v_{(x,\mu)}$. These two parameters control the influence of the fact that a correspondence is also present in the personal best (β) or the neighbourhood best (γ) alignment, respectively. After this, each $v_{(x,\mu)}$ is multiplied by a uniform random number $\phi_\mu = \text{rand}_U \in [0, 1]$. The proportional likelihoods realise the social component, which is typical for PSO algorithms.

To calculate a *keep-set*, *i.e.* a set of correspondences that will *not* be replaced by a random re-initialisation during an iteration, two sets are defined as

$$F_{(x,\kappa)} = \left\{ C_{(x,j_\mu)} \mid \mu \in \{1, \ldots, \kappa \cdot k\}, \ j_\mu \text{ a reordering as in } \vec{F}_x \right\} \tag{5.17}$$

$$V_{(x,\kappa)} = \left\{ C_{(x,l_\mu)} \mid \mu \in \{1, \ldots, \kappa \cdot k\}, \ l_\mu \text{ a reordering as in } \vec{V}_x \right\} \tag{5.18}$$

with a parameter $\kappa \in [0, 1]$ to control the size of the keep-set. The sets $F_{(x,\kappa)}$ and $V_{(x,\kappa)}$ hence contain those correspondences of a particle, which are the $\kappa \cdot k$ best evaluated, and highest ranked according to their proportional likelihood, respectively. The *keep-set* is now defined as

$$K_{(x,\kappa)} = F_{(x,\kappa)} \cap V_{(x,\kappa)} \tag{5.19}$$

containing those correspondences, which are part of both sets $F_{(x,\kappa)}$ and $V_{(x,\kappa)}$. Values for κ should not be chosen too small in order to avoid an empty keep-set after computing the intersection according to Equation (5.19). On the other hand, κ should not be chosen too large either in order to avoid getting stuck in local optima by keeping too large portions of the alignment throughout the iterations. Values for κ of around 0.5 have shown to lead to good results.

For a more stringent convergence towards an optimum alignment, an additional *safe-set* is introduced as

$$S_{(x,\sigma)} = \left\{ C_{(x,j_\mu)} \mid \iota_{(x,\mu)} > 1 - \sigma \right\} \tag{5.20}$$

a set of correspondences, which will never be replaced in this particle. Here $\sigma \in [0, 1]$ is the confidence threshold for correspondences to be included in the safe-set. Since there is the chance of getting stuck in a local optimum for the alignment, one would typically choose a very small value for σ. In each step from iteration i to iteration $(i + 1)$, the update algorithm firstly computes a new particle length k' according to a self-adaptation process as discussed later in this section. Secondly, the particle updates its configuration as

$$\langle I_{i+1},\, p_{i+1} \rangle = U(\langle I_i,\, p_i \rangle) \tag{5.21}$$

where $I_{i+1} = I_i$ and for all $x \in I_{i+1}$

$$p_{i+1}(x) = S_{(x,\sigma)} \cup K_{(x,\kappa)} \cup R \tag{5.22}$$

where R is a set of $k' - |S_{(x,\sigma)} \cup K_{(x,\kappa)}|$ random new correspondences, such that the alignment validity (cf. Definition 2.7) is maintained.

Informally, the particle keeps the set $S_{(x,\sigma)} \cup K_{(x,\kappa)}$ and replaces the remaining $k' - |S_{(x,\sigma)} \cup K_{(x,\kappa)}|$ correspondences with new random ones. This behaviour ensures a convergence of each particle towards an optimum according to the objective function, which is based on Definition 2.9, since the keep-set will steadily increase, and the fluctuation due to random re-initialisation will become less drastic as the swarm evolves.

The presented DPSO differs from the approach by Correa *et al.* mainly in two aspects. Firstly, the size, *i.e.* dimensionality of each particle is updated in each iteration, where in the approach of Correa *et al.* each particle is given a randomly chosen size, which does not change throughout the iterations. In their approach this is reasonable seeing that in their experiment [32] the authors used a population size, which is much larger than the number of possible particle lengths. For the problem of ontology alignment the number of possible particle lengths can be much larger, since it depends on the size of the input ontologies, *i.e.* their number of entities. It might thus become difficult to increase the population size accordingly, which makes it necessary to dynamically adjust the particle lengths in order to find the optimal size of an alignment. The second aspect in which this approach differs from the one by Correa *et al.* is the

Table 5.1.: Example correspondence set and assigned confidence values for a candidate alignment represented by particle x of the two example ontologies presented in Figure 1.1.

Correspondence	Confidence value
$C_{(x,1)} = \langle$Unpublished, Publication$\rangle$	0.67
$C_{(x,2)} = \langle$Inproceedings, InProceedings$\rangle$	0.85
$C_{(x,3)} = \langle$Book, Book$\rangle$	0.96
$C_{(x,4)} = \langle$Mastersthesis, Event$\rangle$	0.13
$C_{(x,5)} = \langle$TechReport, TechnicalReport$\rangle$	0.81

particle update procedure. In this approach, the change of a particle's configuration does not only depend on the configuration of the personal best and neighbourhood best[7], but also on the evaluation of the single correspondences. This is not possible in the use case of attribute selection for a classifier, as attributes cannot be evaluated independently. However, in the presented DPSO, the influence of single correspondence evaluations should not be overemphasised, since the goal is to obtain *globally* good solutions, which have to meet criteria that cannot be expressed by evaluating correspondences in isolation. Thus incorporating the correspondence confidences this way should merely be seen as a heuristic, which has shown to provide good results (cf. Chapter 7).

Example

In order to illustrate the procedure from the previous paragraphs, one iteration is run through in this example, updating a particle x. Consider an alignment of the two example ontologies presented in Figure 1.1. Suppose, x represents an alignment consisting of $k = 5$ correspondences

$$p(x) = \left\{ C_{(x,1)}, C_{(x,2)}, C_{(x,3)}, C_{(x,4)}, C_{(x,5)} \right\}$$

which are allocated as in Table 5.1. Suppose the confidence values of the single correspondences have been determined as follows

$$\vec{F}_x = \begin{pmatrix} 0.96 & 0.85 & 0.81 & 0.67 & 0.13 \\ C_{(x,3)} & C_{(x,2)} & C_{(x,5)} & C_{(x,1)} & C_{(x,4)} \end{pmatrix}$$

---

[7]Correa *et al.* use the entire swarm as neighbourhood, so in their original work they use a *global* best instead of a *neighbourhood* best.

Note that the array is sorted by its confidence values in descending order, as larger values mean a better evaluation.

The velocity vector $\vec{V}_x$ has been initialised with all proportional likelihoods set to 1:

$$\vec{V}_x = \begin{pmatrix} 1 & 1 & 1 & 1 & 1 \\ C_{(x,1)} & C_{(x,2)} & C_{(x,3)} & C_{(x,4)} & C_{(x,5)} \end{pmatrix}$$

Now suppose, correspondences $C_{(x,2)}$, $C_{(x,3)}$, and $C_{(x,5)}$, are also present in $p_{\text{personal}}(x)$, and $C_{(x,3)}$ is also present in $p_{\text{neighbour}}(x)$. Parameters β and γ are added accordingly (*e.g.* $\beta = 0.4$ and $\gamma = 0.5$):

$$\vec{V}_x = \begin{pmatrix} 1 & (1+\beta) & (1+\beta+\gamma) & 1 & (1+\beta) \\ C_{(x,1)} & C_{(x,2)} & C_{(x,3)} & C_{(x,4)} & C_{(x,5)} \end{pmatrix}$$

After adding the parameters, each proportional likelihood is multiplied by a uniform random number $\phi_j = \text{rand}_U \in [0, 1]$, $\forall j \in \{1, \ldots, 5\}$. The array will then be sorted by its proportional likelihoods in descending order, as higher values mean a higher likelihood. This might result in something like

$$\vec{V}_x = \begin{pmatrix} 1.34 & 1.12 & 0.88 & 0.76 & 0.32 \\ C_{(x,2)} & C_{(x,5)} & C_{(x,4)} & C_{(x,3)} & C_{(x,1)} \end{pmatrix}$$

Suppose $\kappa = 0.6$, so the *keep-set* $K_{(x,\kappa)}$ is built as the intersection of the first $\kappa \cdot k = 3$ correspondences of the arrays $\vec{F}_x$ and $\vec{V}_x$, which results in

$$K_{(x,\kappa)} = \{C_{(x,2)}, C_{(x,5)}\}$$

Assume that $\sigma = 0.1$ and thus the *safe-set* $S_{(x,\sigma)}$ determined as

$$S_{(x,\sigma)} = \{C_{(x,3)}\}$$

The update algorithm will now keep the set

$$S_{(x,\sigma)} \cup K_{(x,\kappa)} = \{C_{(x,2)}, C_{(x,3)}, C_{(x,5)}\}$$

and replaces the remaining two correspondences with random new ones.

Self-Adaptation of Particle Length

A general problem when aligning two ontologies is that the optimal number of correspondences is not known upfront. This method approaches this by assigning each particle a random number of correspondences during the initialisation. The initial guesses are uniformly distributed between zero and the maximum number of possible correspondences between the two ontologies. Assuming that the chances for a particle to receive a good fitness value are higher if its number of correspondences is close to the optimal number of correspondences, the heuristic below attempts to adjust the number of correspondences for each particle and in each iteration based on the current neighbourhood best particle. Let k_x be the number of correspondences represented by particle x, and let k_{nBest} be the number of correspondences represented by the *nBest*. (Note that correspondences for each entity type are considered separately.) Each particle adjusts its number of correspondences if the following expression becomes true:

$$\begin{cases} r_1 \geq \tau_i & \text{if } k_{\text{nBest}} > k_x \\ r_1, r_2 \geq \tau_i & \text{if } k_{\text{nBest}} < k_x \\ \text{false} & \text{else} \end{cases} \tag{5.23}$$

where $r_1 = \text{rand}_U$ and $r_2 = \text{rand}_U$ denote random values and τ_i an iteration dependent threshold value defined as

$$\tau_i = \lambda \left(\frac{i}{i_{max}} \right)^2 \tag{5.24}$$

where $\lambda \in [0, 1]$ is a constant weighting factor and i and i_{max} denote the current and maximum iteration, respectively. The probability for a change therefore increases with the number of iterations which prevents very rapid changes of the number of correspondences at the beginning of the process where the prediction of the actual number of correspondences is less accurate than later in the optimisation. Furthermore the probability for decrease is always significantly lower than for an increase. The underlying assumption is that more correspondences are generally more desirable and this scheme has proven to be successful.

For each of the particles whose number of correspondences are to be changed, the maximum range of this change is determined by

$$\Delta k_x^{max} = \begin{cases} w_{inc} \cdot (k_{\text{nBest}} - k_x) & \text{if } k_{\text{nBest}} > k_x \\ w_{dec} \cdot (k_x - k_{\text{nBest}}) & \text{if } k_{\text{nBest}} < k_x \end{cases} \tag{5.25}$$

where w_{inc} and w_{dec} denote weighting factors for the size of the interval. The extended range ensures, similar to the velocity update in continuous PSO that the interval exceeds the distance between the two values and allows a new value to either under- or overshoot the reference value, *i.e.* the number of correspondences of the neighbourhood best particle. The new number of correspondences of each type for a particle x is then adjusted by a random value $\Delta k_x = \mathrm{rand}_U(0, \ldots, \Delta k_x^{max})$. The new number of correspondences k'_x of particle x can be computed as

$$k'_x = \begin{cases} k_x + \Delta k_x & \text{if } k_{\mathrm{nBest}} > k_x \\ k_x - \Delta k_x & \text{if } k_{\mathrm{nBest}} < k_x \end{cases} \tag{5.26}$$

In the case of an increase the algorithm attempts to adopt these from the keep-set of the neighbourhood best particle. If more new correspondences are needed than can be added this way the remaining correspondences are randomly created. In either case only valid correspondences are added, *i.e.* the new correspondences cannot violate constraints such as the restriction to "?:?" alignments, *etc.* When on the other hand the number of correspondences decreases, a fitness ranking of all correspondences is performed and the worst performing elements are removed.

Algorithm

This subsection presents an algorithm that computes an ontology alignment following the method presented in the previous paragraphs. In this presentation the algorithm is split into three parts, an initialisation step, the swarm iteration, and an update procedure to determine the new configuration of each particle.

The computation of an alignment starts with an initialisation, encoded in Algorithm 5.5. In this initialisation step, each particle is initialised with a random number of correspondences. It also encompasses evaluation, *i.e.* computation of the fitness value of each correspondence and the initial assertion of the personal best alignment.

The execution of the algorithm is an iterative, guided evolution of the particle swarm as outlined in Algorithm 5.6. In each iteration, the personal and neighbourhood best alignment is updated, if a new best performing particle is seen. The guided evolution of particles behaves according to the update procedure denoted by Algorithm 5.7. Note that each particle can be evaluated and updated in parallel.

The particle update procedure in Algorithm 5.7 states the formal definitions presented earlier in this section in a sequential manner. The single steps are explained in detail there.

Algorithm 5.5 Initialisation of Particles

Require: $|I|$ the number of particles

> **for** $i = 1$ **to** $|I|$ **do**
>> **for all** $t \in \mathcal{T}$ **do**
>>> $n_t = \min\{\sharp_t \mathcal{O}_1, \sharp_t \mathcal{O}_2\}$
>>> $k_t \Leftarrow \mathrm{rand}_U(\{1, \ldots, n_t\})$, a uniform random number
>>> **for** $j = 1$ **to** k_t **do**
>>>> Randomly select entities $e_1 \in \mathrm{voc}_t(\mathcal{O}_1)$ and $e_2 \in \mathrm{voc}_t(\mathcal{O}_2)$
>>>> that have not already been selected,
>>>> and create correspondence $C_j = \langle e_1, e_2 \rangle$
>>>> Compute $\iota(C_j)$
>>>> $p(x_i) \Leftarrow p(x_i) \cup \{C_j\}$
>>> **end for**
>>> Build $\vec{F}_i$ according to (5.15)
>>> Compute $F(p(x_i))$
>>> $p_{\mathrm{personal}}(x_i) \Leftarrow p(x_i)$
>> **end for**
> **end for**

5.4. Discussion

This section concludes the chapter by providing a discussion of the presented algorithms on the theoretical level. The discussion focuses on two aspects: *(i)* the crossover and mutation operators, and the reason why the former is not used in the presented Evolutionary Algorithm, and *(ii)* the differences and similarities between the Evolutionary Algorithm and the Particle Swarm Optimisation algorithm.

5.4.1. Mutation vs. Crossover

As part of the update operation in Genetic Algorithms a recombination (crossover) is performed, where two parent individuals are selected and sections of their solution strings are exchanged. The motivation of this operation is the idea that exchanging partial information from two good solutions bears the chance of obtaining a better solution. While this idea of recombination straightforwardly applies to simple solution representations, there are difficulties when permutation-like representations are

---

Algorithm 5.6 Swarm Iteration

Require: $|I|$ the number of particles,
  i_{max} the number of iterations
    for $i = 1$ **to** i_{max} **do**
      for $j = 1$ **to** $|I|$ **do**
        Update particle $x_j \in I$ according to Algorithm 5.7
        if $F(p(x_j)) > F(p_{\text{personal}}(x_j))$ **then**
          $p_{\text{personal}}(x_j) \Leftarrow p(x_j)$
        end if
        if $F(p(x_j)) > F(p_{\text{neighbour}}(x_j))$ **then**
          $p_{\text{neighbour}}(x_j) \Leftarrow p(x_j)$
        end if
      end for
    end for

---

used. Since applying the crossover operation for two permutations typically generates offsprings which are not valid permutations anymore, corrective measures have been proposed [132]. Such measures are reasonable if they preserve crucial information that are a determining factor for the quality of the parent solution participating in the crossover.

Example. Consider a candidate solution for the Travelling Salesman Problem (TSP) that is represented by a permutation. The *order* in which cities are visited is an important factor for the solution quality. Hence, a corrective measure for permutation recombination that reconstructs a valid permutation string from the parental fragments while preserving the relative order of their elements is desired.

A correspondence permutation as from Definition 5.1 does not have the ordering as decisive factor but the assignment of values to specific positions. Hence, corrective measures that change the exact position of a value in the correspondence permutation lose valuable parental information if that position was decisive for the quality of the parent. The GAOM approach for ontology alignment by Wang *et al.* [130] does apply crossover, but does not perform any corrective measures. This causes the undesirable limitations studied in Section 3.2.1, such as enforced coverage of one ontology and enforced 1:m alignment cardinality.

There has been continuous argument between advocates of the Genetic Algorithm community and promoters of Evolutionary Programming re-

---

Algorithm 5.7 Update of Particle x_i

Require: k the number of correspondences in this particle
  $\beta, \gamma, \kappa, \sigma$ parameters
  $\vec{V_i}$ the proportional likelihood vector
  $\vec{F_i}$ the evaluation vector

  Compute new particle length k' according to (5.26)
  (this also modifies x_i by adding or removing correspondences according to the length adjustment)
  for $\mu = 1$ **to** k' **do**
      if $C_{(i,l_\mu)} \in p_{\text{personal}}(x_i)$ **then**
          $v_{(i,\mu)} \Leftarrow v_{(i,\mu)} + \beta$
      end if
      if $C_{(i,l_\mu)} \in p_{\text{neighbour}}(x_i)$ **then**
          $v_{(i,\mu)} \Leftarrow v_{(i,\mu)} + \gamma$
      end if
      $v_{(i,\mu)} \Leftarrow v_{(i,\mu)} \cdot \phi_\mu, \quad \phi_\mu = \text{rand}_U \in [0, 1]$ a uniform random number
  end for
  Sort $\vec{V_i}$ by v_i in descending order
  Sort $\vec{F_i}$ by f_i in descending order
  Compute $K_{(i,\kappa)}$ according to (5.17), (5.18), and (5.19)
  Compute $S_{(i,\sigma)}$ according to (5.20)
  Replace correspondences $p(x_i) \setminus (S_{(i,\sigma)} \cup K_{(i,\kappa)})$ by the same number of randomly generated new ones
  for all $C \in p(x_i)$ **do**
      Compute $\iota(C)$
  end for
  Compute $F(p(x_i))$

---

garding usefulness and necessity of the mutation and crossover operators [118]. Genetic Algorithms employ the crossover operator, whereas Evolutionary Programming deliberately abandons crossover in favour of the exclusive application of mutation. It is generally accepted that crossover is best suitable for *exploitation* of areas of the problem space where there are known good solutions, while mutation is better for *exploration* of new areas in the problem space [118]. As discussed in the previous paragraphs, the exploitation behaviour of crossover cannot easily be achieved for the correspondence permutation in ontology alignment.

Additionally, mutation is considered important in non-stationary environments [118], which paves the way for applying the presented Evolutionary Algorithm for ontology alignment also for scenarios with changing ontologies.

5.4.2. Evolutionary Algorithm vs. Particle Swarm Optimisation

Evolutionary Algorithms and *Particle Swarm Optimisation* as two representative biologically-inspired optimisation paradigms have been chosen for implementing prototypical algorithms to solve the ontology alignment problem. Both approaches are population-based and thus could be formalised in a similar fashion using the generic notations introduced in Section 2.3. The commonalities become clear in the similar structure of the algorithms, which could both be described in three analogous parts of initialisation (Algorithms 5.1 and 5.5), population iteration (Algorithms 5.2 and 5.6), and individual update (Algorithms 5.3 and 5.7).

Objective Function. Both algorithms use the same objective function F for evaluating alignments, and a support heuristic ι for evaluating correspondences. They use both objective function and support heuristic as a black box, *i.e.* they do not depend on their internal structure. This independence underpins the hypothesis from the introductory Conjecture 2 that biologically-inspired optimisation techniques can be applied for ontology alignment despite different ontology characteristics. These characteristics can be encoded into the objective function using the extensible "toolbox" of similarity metrics and aggregation functions presented in Chapter 4. Furthermore, the global alignment evaluation performed by F accounts for global metrics, substantiating the introductory Conjecture 3. This holds for both Evolutionary Algorithms as well as Particle Swarm Optimisation.

Support Heuristic. Both algorithms use a support heuristic ι reflecting the confidence of any correspondence participating in the alignment. The use of this support heuristic is based on the assumption that the quality of an alignment is to a significant extent governed by the quality of its correspondences. This makes the application of biologically-inspired optimisation techniques and the encodings presented in this chapter less an "uninformed" search as it is typically the case in application domains of those metaheuristics. On the other hand exploiting the correspondence level information in the update operations of the algorithm seems a reasonable approach to foster convergence.

In the case of the Evolutionary Algorithm, the mutation operators make use of the correspondence confidence values when choosing a correspondence to be altered. In the case of Particle Swarm Optimisation, the keep-set for each particle is partially, and the safe-set entirely determined by the confidence values of represented correspondences.

Social Component.   A major difference between the two paradigms, Evolutionary Computation and Swarm Intelligence, is the social component being exploited in the latter. While in the Evolutionary Algorithm global alignment criteria are mostly considered by favouring well performing solutions during the selection process, this information is communicated via a social network in the presented Particle Swarm Optimisation algorithm.

Compared to the selection operation in the Evolutionary Algorithm, the propagation via the social network in the Particle Swarm Optimisation approach is expected to have less drastic influence, since it does not replace or remove solutions, but "only" changes probabilities.

6. Implementation

This algorithms presented earlier in this book are implemented as several prototypes that have been developed in the context of the *THESEUS* programme[1] funded by the German Ministry of Economics and Technology. Within the *THESEUS* programme, the work package "Ontology Management" of the Core Technology Cluster (CTC) has been responsible for providing a scalable ontology management infrastructure to be applied to various use case domains. The prototypes presented in this chapter constitute the *THESEUS* component *HARMONIA*.

On the one hand the implementation serves the purpose of enabling the evaluation of the concepts introduced in Chapter 5. On the other hand, it is intended to provide software artifacts that are ready to be used and extended, and thus enable the research community to pick up ideas and extend them in various directions. In order to facilitate the latter, several design decisions were taken and certain software development techniques were applied.

The pursued implementation objectives can be summarised as follows:

1. *Reusability*: Decoupling of independent or shared software modules enables reusability of components in other components and applications.

2. *Robustness*: Object-oriented software development and design patterns increase stability, efficiency, and testability.

3. *Flexibility*: The large amount of parameters that influence the behaviour of the implemented approaches requires means to adjust the configuration without major efforts. To achieve this all important parameters can be set via external configuration files.

All components have been implemented in the Java™ 6 programming language due to its wide acceptance in the community of semantic tech-

---

[1] `http://theseus-programm.de/` (accessed March 22, 2012)

nology researchers and the availability of powerful 3rd-party libraries, such as the OWL API.

This chapter describes four software components. An application programming interface (API) for ontology alignments named *KADMOS* is presented in Section 6.1. Implementations of the evaluation metrics for correspondences and alignments as discussed in Chapter 4 are presented in Section 6.2. The Sections 6.3 and 6.4 describe the specific implementations for the Evolutionary Algorithm approach (MapEVO) and the Particle Swarm Optimisation approach (MapPSO) for ontology alignment, respectively. Section 6.5 presents several ways of how the algorithms can be deployed and thus made usable by client applications. Apart from deployment as application programming interface (API), Web Service, and a proprietary packaging format for an evaluation platform, the section focuses on deployment in a cloud computing infrastructure in order to exploit the parallel nature of the population-based optimisation algorithms. Finally, in Section 6.6 some notes about the development infrastructure are presented.

6.1. *KADMOS* API

Any alignment algorithm requires efficient means to represent and operate on alignments. There is an open-source implementation project of an Application Programming Interface (API) for ontology alignment[2] [37], initiated by INRIA Grenoble Rhône-Alpes[3]. This API is frequently used for developing alignment systems, however, it has several shortcomings. Firstly, there is a negligent use of object typing. The static typing mechanism of the Java™ programming language is misused by defining general `java.lang.Object` argument types in core methods in order to pretend flexibility regarding the ontology representation framework. This, in accordance with poor documentation, hampers ease of use and increases the chance of runtime errors. Secondly, the API does not allow for a clear separation of concerns regarding alignment representation and algorithm implementation. Alignment algorithms implementing the IN-

---

[2]`http://alignapi.gforge.inria.fr/`
(accessed November 28, 2011)
In the remainder of this chapter, this API is called "INRIA Alignment API" in order to avoid confusion with the *KADMOS* API.
[3]`http://www.inria.fr/centre/grenoble`
(accessed November 28, 2011)

RIA Alignment API's `AlignmentProcess` interface have to represent the complete alignment themselves. Thirdly, API objects are not serialisable, which hinders the implementation of distributed alignment algorithms that need to communicate by sending and receiving API objects.

These shortcomings gave reason to implement a novel simple, efficient alignment API as an alternative to be used in various alignment algorithms. This API, named *KADMOS* API, is more restricted compared to the INRIA Alignment API in terms of functionality, since it is being developed from scratch. However, it overcomes the issues mentioned above by Java™ generics in order to guarantee correct API usage, separation of representation and algorithm, and serialisation facilities.

The *KADMOS* API uses the OWL API[4] [67] in order to represent OWL ontology objects.

6.1.1. Core Representation API

The core of the *KADMOS* API is the representation of correspondences and alignments. Figure 6.1 shows the interfaces and methods they expose. The generic `Correspondence` interface represents a correspondence of two entities of the same type T, which has to be a subtype of `OWLEntity` provided by the OWL API. In addition to the two entities, a correspondence has a confidence value to be set by algorithms in order to make a statement about how certain the correspondence is. The interface `Alignment` represents an alignment and provides access to the correspondences it contains. Most of its methods are for convenience and allow for implementing classes to use various index data structures in order to efficiently access correspondences. Both `Correspondence` and `Alignment` implement the marker interface `Evaluable` in order to allow static typing in other modules using the *KADMOS* API for assessing correspondences or alignments, such as the *HARMONIA* Commons module (cf. Section 6.2).

For efficient instance control and correct instantiation, the *factory* design pattern is used for creating correspondences and alignments. Every implementing class is expected to also provide a specific factory for instantiating it. This mechanism is used for instance in the `AlignmentParser` that uses the specified factories to create an object model of an externalised alignment, *e.g.* an XML-based file representation. Analogously,

---

[4]`http://owlapi.sourceforge.net/`
(accessed March 22, 2012)

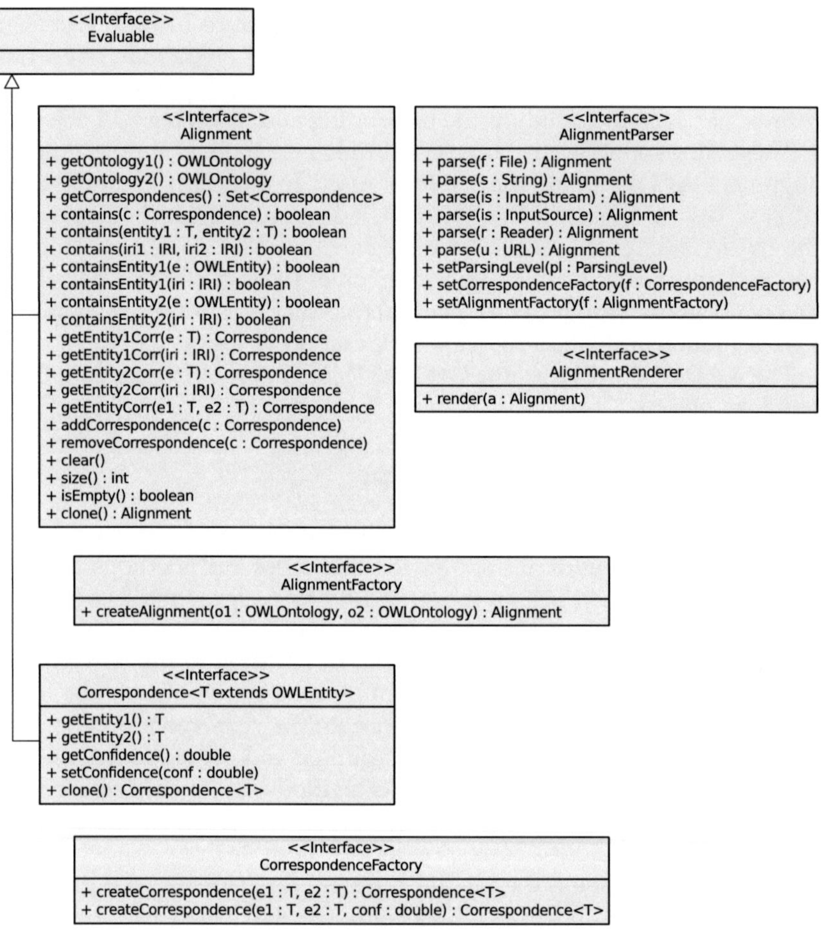

Figure 6.1.: UML class diagram of the *KADMOS* API.

an `AlignmentRenderer` interface is provided to allow for different implementations for externalising alignments.

6.1.2. Alignment Algorithm API

Decoupled from the representation API for correspondences and alignments, an algorithm API is provided as part of the *KADMOS* API. Figure 6.6 illustrates two interfaces to be implemented by alignment algorithms. Algorithms that take two ontologies and compute an alignment can implement the `AlignmentAlgorithm` interface. Algorithms that are capable of using an initial alignment, *i.e.* an alignment or partial alignment that is known prior to the execution of the algorithm, can implement the `UpdatingAlignmentAlgorithm` interface. The algorithm can be provided with arbitrary configuration parameters via an optional method `setParameters`.

6.1.3. Cloud Adapter API

In order to facilitate the use of cloud infrastructures the *KADMOS* API provides interfaces and implementations for connecting to computing resources of "Infrastructure-as-a-Service" (IaaS) providers. Figure 6.2 illustrates the cloud adapter API included in *KADMOS*. It includes two generic interfaces, `CloudController` and `CloudWorker`, that can be used by algorithms to follow the common server/worker pattern. These interfaces can be implemented for any IaaS cloud computing environment. There are implementations provided by *KADMOS* for the Amazon Web Services[TM] (AWS) infrastructure, where the management of Amazon Machine Images (AMIs) is encapsulated by an `AWSAdapter`. Since computationally intensive algorithms, such as ontology alignment algorithms, can have long response times for single worker jobs, it is crucial for the controlling instance to know whether a worker is still alive and computing, or whether it is no longer responsive due to a network problem or the like. Hence, *KADMOS* provides the `ServerHeartbeat-Communicator` and `ClientHeartbeatCommunicator` facilities in order to send a periodic heartbeat in case a worker response is taking a long time. Messages sent between server and worker, as well as the heartbeat signal, are encapsulated in a `SimpleCloudMessage` object.

6.2. *HARMONIA* Commons

Shared components available to be used by arbitrary alignment algorithms have been implemented in the module *HARMONIA* Commons. The module is decoupled from any particular alignment algorithm and uses the *KADMOS* API for representing alignment objects. Essentially this component can be used to assess any *KADMOS* object that is charac-

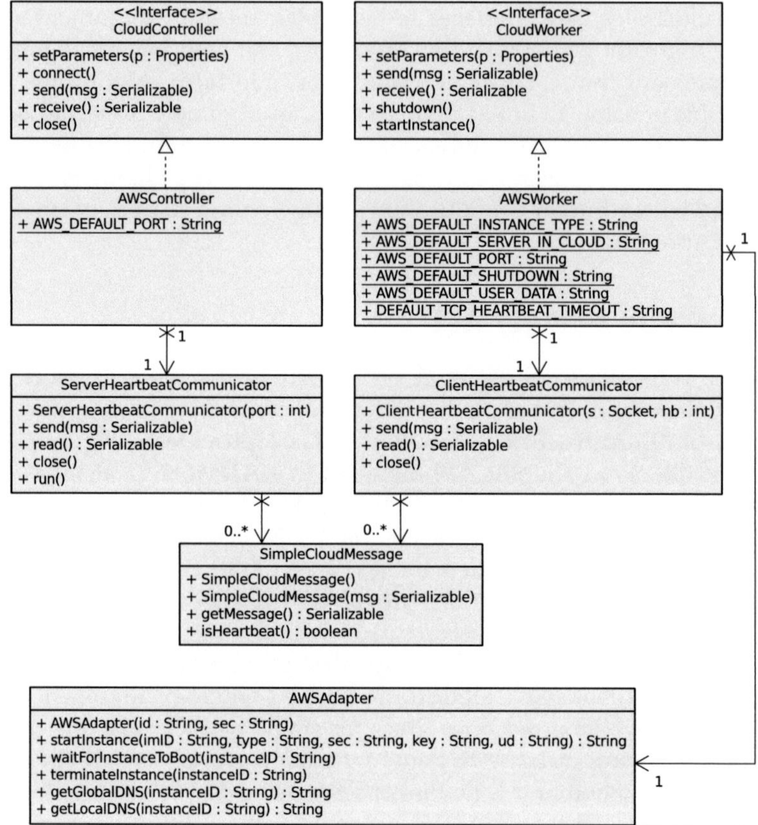

Figure 6.2.: UML class diagram of the *KADMOS* cloud adapter API.

terised as `Evaluable` by the according marker interface. Currently these objects are correspondences and alignments.

The design of the module defines a generic interface `Evaluator` with three specialisations:

- Correspondence evaluation (`BaseMatcher`)

- Alignment evaluation (`AlignmentEvaluator`)

- Evaluation aggregation (`Aggregator`)

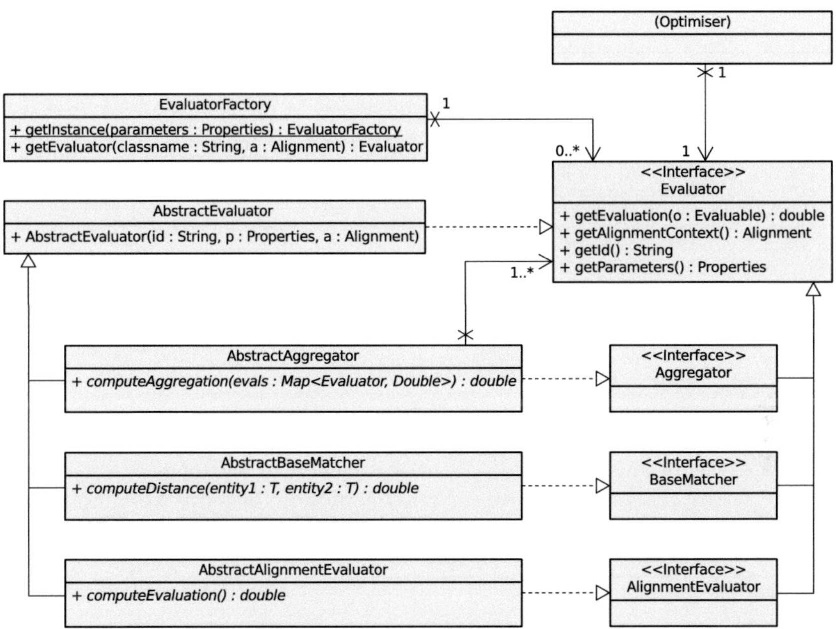

Figure 6.3.: UML class diagram of the *HARMONIA* Commons module for alignment evaluation.

Correspondence evaluators and alignment evaluators are only applicable to the types `Correspondence` and `Alignment`, respectively, while aggregators can be applied to any evaluable object. Aggregators have several other evaluators registered and aggregate their individual evaluations to a single evaluation value. The generic availability of aggregators allows for the configuration of complex alignment evaluation functions that comprise global measures on the alignment level, as well as local measures on the correspondence level.

Figure 6.3 shows the UML class diagram for the alignment evaluation module of *HARMONIA* Commons. It illustrates, how abstract classes implement core functionality for each evaluator type. Concrete implementations thus have to extend these abstract classes and have to implement a single abstract method. The class indicated as (`Optimiser`) in the diagram represents any of the optimisation metaheuristics investigated in this book. However, any ontology alignment algorithm (re-

Table 6.1.: Implementation classes for local correspondence evaluators. Package names are omitted in order to improve readability. (All the local correspondence evaluators are implemented in the *HARMONIA* package `de.fzi.harmonia.commons.basematcher.*`.)

Implementation class	Evaluation function
`EntityNameDistanceMatcher`	$h_{\text{lexIDSim}}(C)$
`EntityLabelDistanceMatcher`	$h_{\text{lexLabelSim}}(C)$
`EntityTextNameDistanceMatcher`	$h_{\text{textIDSim}}(C)$
`EntityTextLabelDistanceMatcher`	$h_{\text{textLabelSim}}(C)$
`EntityCommentDistanceMatcher`	$h_{\text{entityCommentSim}}(C)$
`EntityVirtualDocumentDistanceMatcher`	$h_{\text{entityVDSim}}(C)$

Table 6.2.: List of implementation classes for contextual correspondence evaluators. Package names are omitted in order to improve readability. (All contextual correspondence evaluators are implemented in the *HARMONIA* package `de.fzi.harmonia.commons.basematcher.*`.)

Implementation class	Evaluation function
`HierarchyDistanceMatcher`	$h^A_{\text{hierarchy}}(C)$
`SimilarityFloodingDistanceMatcher`	$h^A_{\text{hierarchyProp}}(C)$
`ClassAsDROfPropertyMatcher`	$h^A_{\text{classDRProp}}(C)$
`PropertyByDRClassMatcher`	$h^A_{\text{propDRClass}}(C)$
`CrissCrossMatcher`	$h^A_{\text{crissCross}}(C)$
`SimpleBlackBoxExplanationMatcher`	$h^A_{\text{explanation}}(C)$

gardless whether it is optimisation-based or not) can use the module to evaluate alignments or correspondences.

The *HARMONIA* Commons evaluation module contains implementation classes for the evaluation functions defined in Section 4.1. This is by no means an exhaustive set of implemented evaluation measures and can be extended according to whatever alignment quality criteria are relevant for the alignment task at hand. There is no clear separation between local and contextual correspondence evaluators on the implementation side, since the alignment context has to be provided at initialisation time for all evaluators. Table 6.1 lists the implementation classes of local correspondence evaluators for the evaluation functions defined in Section 4.1.1. Table 6.2 lists the implementation classes of contextual correspondence

Table 6.3.: List of implementation classes for global alignment evaluators. Package names are omitted for readability purposes. (All the global alignment evaluators are implemented in the *HARMONIA* package `de.fzi.harmonia.commons.alignmentevaluator.*`)

Implementation class	Evaluation function
`CorrespondenceEvaluator`	$H_{\mathrm{corrContrib}}(A)$
`AlignmentSizeEvaluator`	$H_{\mathrm{size}}(A)$
`AlignmentConsistencyEvaluator`	$H_{\mathrm{consist}}(A)$
`AlignmentCoherenceEvaluator`	$H_{\mathrm{coherence}}(A)$
`StructuralPreservationEvaluator`	$H_{\mathrm{structPreserv}}(A)$
`CrissCrossAlignmentEvaluator`	$H_{\mathrm{crissCross}}(A)$

Table 6.4.: Implementation classes for aggregators. Names of packages are omitted due to readability. (All aggregators are implemented in the package `de.fzi.harmonia.commons.aggregator.`)

Implementation class	Aggregation function
`MaxAggregator`	$\Gamma_{\mathrm{max}}(\vec{f}, \vec{\omega})$
`WeightedAverageAggregator`	$\Gamma_{\mathrm{weightAvg}}(\vec{f}, \vec{\omega})$
`OWAAggregator`	$\Gamma_{\mathrm{owa}}(\vec{f}, \vec{\omega})$

evaluators for the evaluation functions defined in Section 4.1.2. The implementation classes for alignment evaluators are listed in Table 6.3 along with the according evaluation functions defined in Section 4.1.3. The implementation classes for aggregators are listed in Table 6.4 along with the according aggregation functions defined in Section 4.2.

The setup of the complete evaluation function is configured via a common parameter object that is passed to every evaluator when it is created. Every evaluator is assigned a unique identifier (specified in the configuration parameters), which allows for using evaluators of the same type in different places of the evaluation function. To this end, the direct "user" of an evaluator is responsible for its initialisation, *e.g.* an aggregator is responsible for the initialisation of the evaluators, whose values it aggregates. In order to guarantee consistent and efficient instance creation, an `EvaluatorFactory` is provided. This factory implements instance control mechanisms in order to avoid too many redundant objects. To this end, there can be only one factory instance for each set of configuration

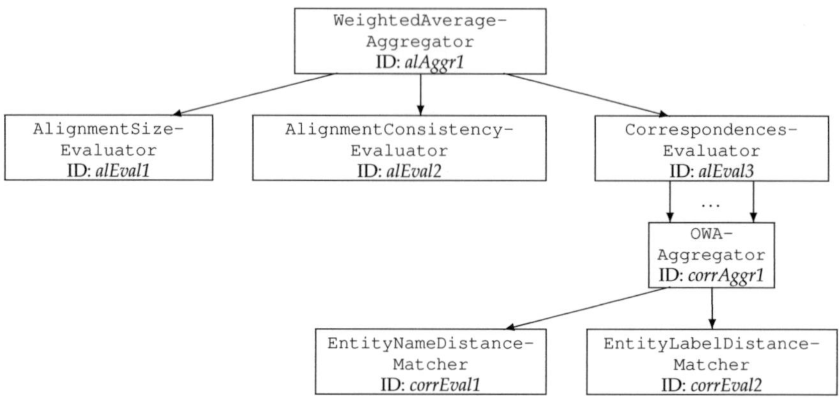

Figure 6.4.: Example implementation of an objective function for evaluating a single alignment using *HARMONIA*.

parameters. Further, the factory creates only a single instance of an evaluator for every evaluator type (specified by its class name) with a given identifier, and for a given alignment context. With particular respect to population-based optimisation algorithms, where each individual of the population represents a valid alignment, there is one alignment context for each individual.

Example. The scenario for single-objective optimisation-based alignment approaches as presented in this book requires the computation of a single fitness value for a complete alignment in each iteration. This fitness value has to comprise global alignment assessments at the alignment level, as well as local assessments at the correspondence level. The fitness function is computed in several stages. In a *first stage* the alignment algorithm instantiates an aggregator that combines various global alignment assessments. As an example let this aggregator be a `Weighted-AverageAggregator` for the several single (global) alignment evaluators, here the `AlignmentSizeEvaluator`, `AlignmentConsistency-Evaluator`, and the `CorrespondencesEvaluator`. These alignment evaluators form the *second stage*. The `CorrespondencesEvaluator` is a special case in the sense that is an alignment evaluator, which acts itself as an aggregator in order to compute a single value from the evaluations of all correspondences in the alignment. Since there are different similarity metrics that can be applied to evaluate a correspondence (implemented as `BaseMatcher` objects), those values must themselves be aggregated into a single evaluation for each correspondence. This correspondence level aggregation is the *third stage*. Here, for instance, an `OWAAggregator` can be applied to favour better evaluation scores. The single base matchers providing similarity measures for individual correspondences are the *fourth stage* in the composition hierarchy. Figure 6.4 illustrates this example.

Note that the described setup is not hard-coded, but fully configured according the single parameter object externally provided. The listing of such an external parameter file for this example is shown in Figure 6.5.

6.3. MapEVO

The MapEVO system is a prototypical implementation of the Evolutionary Algorithm for ontology alignment that was formally introduced in Section 5.3.1. It utilises the *KADMOS* API and the *HARMONIA* Commons library presented in the preceding Sections 6.1 and 6.2, for representing and evaluating alignments and correspondences. The central MapEVO entry point is the class

```
de.fzi.mapevo.algorithm.MapEVOAlignmentAlgorithm
```

that implements the *KADMOS* interface

```
de.fzi.kadmos.api.algorithm.AlignmentAlgorithm
```

as illustrated in Figure 6.6. Apart from this top-level entry point, the implementation of MapEVO comprises objects representing species as population members, the correspondence permutation data structure (cf. Section 5.2.2), as well as encapsulations of the mutation operators (cf. Section 5.3.1).

Omitting the implementation details at the level of utilities and data structure objects, Figure 6.7 shows a sequence diagram of the core components of the MapEVO system. The diagram illustrates the sequence of interactions between system objects in the course of one alignment request, which is triggered by an external caller that is invoking the `align` method of the `MapEVOAlignmentAlgorithm` object. This object subsequently initialises a set of `Species`, which themselves initialise two `Evaluator` instances for the evaluation of correspondences and complete alignments, respectively, as well as the `MutationOperator` object.

```
1   <?xml version="1.0" encoding="UTF-8"?>
2   <!DOCTYPE properties SYSTEM
3       "http://java.sun.com/dtd/properties.dtd">
4
5   <properties>
6   <comment>
7     Sample configuration parameters file for dynamic
8     evaluator configuration.
9   </comment>
10  <!-- ... -->
11
12  <entry key="globalEvaluator">
13    de.fzi.harmonia.commons.aggregator.WeightedAverageAggregator
14      globalEvaluatorID
15  </entry>
16  <entry key="globalEvaluatorID.evaluators">
17    de.fzi.harmonia.commons.alignmentevaluator.\
18  AlignmentSizeEvaluator
19      alEval1
20    de.fzi.harmonia.commons.alignmentevaluator.\
21  AlignmentConsistencyEvaluator
22      alEval2
23    de.fzi.harmonia.commons.alignmentevaluator.\
24  CorrespondenceEvaluator
25      alEval3
26  </entry>
27  <entry key="globalEvaluatorID.weights">
28    0.3 0.3 0.4
29  </entry>
30  <entry key="alEval3.evaluator">
31    de.fzi.harmonia.commons.aggregator.OWAAggregator corrAggr1
32  </entry>
33  <entry key="corrAggr1.evaluators">
34    de.fzi.harmonia.commons.basematcher.EntityNameDistanceMatcher
35      corrEval1
36    de.fzi.harmonia.commons.basematcher.EntityLabelDistanceMatcher
37      corrEval2
38  </entry>
39  <entry key="corrAggr1.weights">
40    0.6 0.4
41  </entry>
42
43  <!-- ... -->
44  </properties>
```

Figure 6.5.: Excerpt of an example configuration parameters file (in the XML serialisation format of `java.util.Properties`).

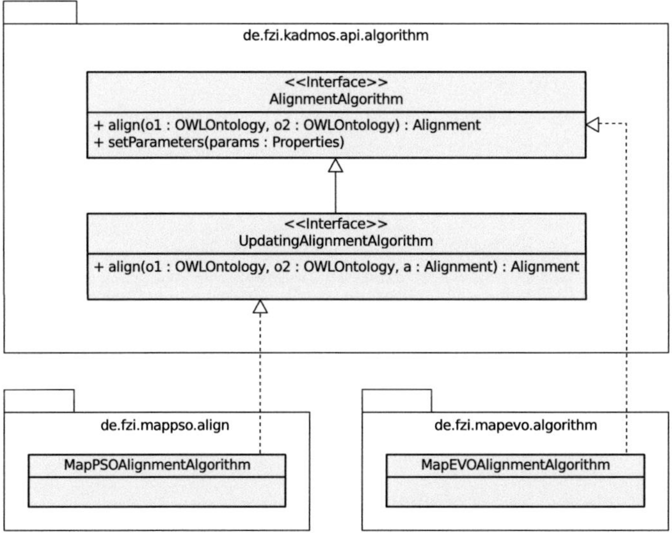

Figure 6.6.: UML class diagram for the MapEVO and MapPSO implementations of the *KADMOS* alignment algorithm interfaces.

For each of the fixed number of selection steps to be performed, the algorithm runs an according fraction of the total number of iterations. In each iteration, the following steps are performed:

1. Swap operation (probability determined as formally described in Section 5.3.1)

2. Evaluation of those correspondences that changed in the previous mutation operation.

3. Exchange operation (probability determined as formally described in Section 5.3.1)

4. Evaluation of those correspondences that changed in the previous mutation operation.

5. Evaluation of the complete alignment represented by the new state of the correspondence permutations.

In the selection step the population is updated according to the selection method described in Section 5.3.1.

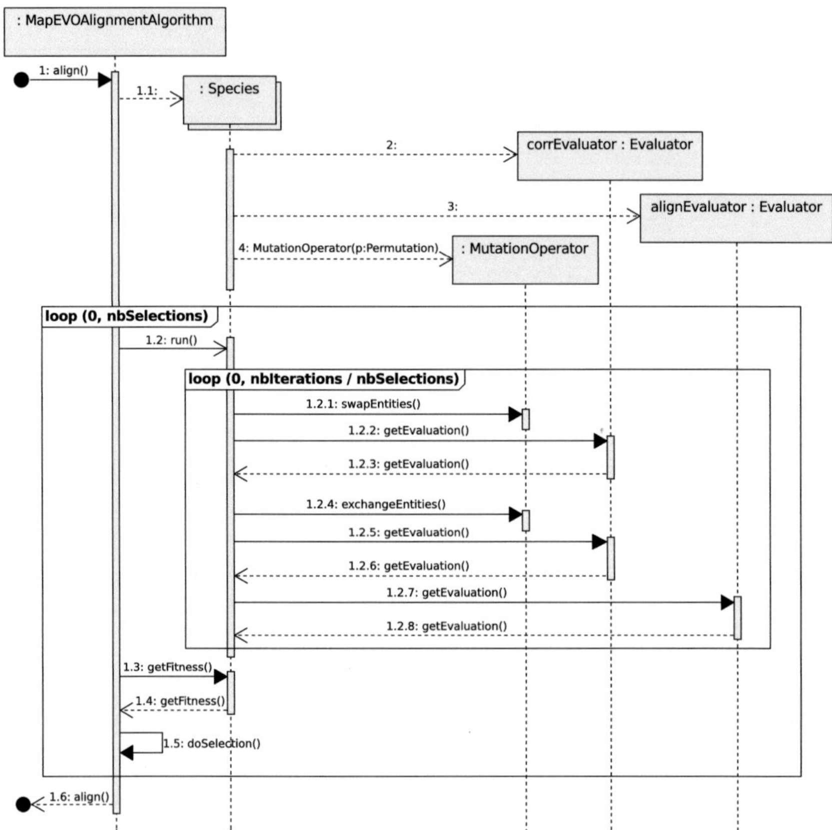

Figure 6.7.: UML sequence diagram of the MapEVO algorithm.

6.4. MapPSO

The MapPSO system is a prototypical implementation of the algorithm
for ontology alignment based on Discrete Particle Swarm Optimisation,
formally introduced in Section 5.3.2. It utilises the *KADMOS* API and the
HARMONIA Commons library presented in the preceding Sections 6.1
and 6.2 for representing and evaluating alignments and correspondences.
The central MapPSO entry point is the class

```
de.fzi.mappso.align.MapPSOAlignmentAlgorithm
```

that implements the *KADMOS* interface

```
de.fzi.kadmos.api.algorithm.UpdatingAlignmentAlgorithm
```

as illustrated in Figure 6.6. Apart from this top-level entry point, the implementation of MapPSO comprises objects representing particles as population members, as well as particle clusters managing collections of particles. The concept of particle clusters allows for the agglomeration of a number of particles in one compute node. In its basic configuration a particle cluster contains one single particle, which reflects the standard setting in Particle Swarm Optimisation. However, putting more than one particle into a cluster allows for *particle pooling*, which is a purely practical solution to address problems occurring in distributed computing environments, such as cloud infrastructures (cf. Section 6.5.3). The social component of Particle Swarm Optimisation algorithms additionally requires the implementation of neighbourhoods and communicators, particularly considering the multi-threaded execution in a potentially distributed computing environment.

Figure 6.8 shows a sequence diagram of the core components of the MapPSO system. The diagram illustrates the sequence of interactions between system objects in the course of one alignment request, which is triggered by an external caller that is invoking the `align` method of the `MapPSOAlignmentAlgorithm` object. The algorithm subsequently initialises a set of `ParticleCluster` objects, a `Topology`, which describes the social network structure between particle clusters, and a `Cluster-Communicator`, which is responsible for propagating the new best alignment being found in any particle cluster. Several implementations have been provided realising concrete variants of particle clusters, social network topologies, and communicators, which are listed in Table 6.5.

The sequence diagram in Figure 6.8 shows a more complex interaction between objects than the corresponding diagram for MapEVO (Figure 6.7) due to the implementation of the social component. To this end, each `ParticleCluster` manages a set of `AlignmentParticle` objects. Invoking the `update` method of an `AlignmentParticle` triggers the "relocation" of the particle in the problem space as formally described in Section 5.3.2 (the sequence diagram omits the details on this level). In order to perform this "relocation", each particle's fitness must be evaluated by calling the `Evaluator`. Each `ParticleCluster` obtains all personal best configurations of its `AlignmentParticle` objects and, in case a new best alignment is found in any of them, communicates this new best alignment to the `ClusterCommunicator`. The `ClusterCommunicator` uses the `Topology` in order to propagate the

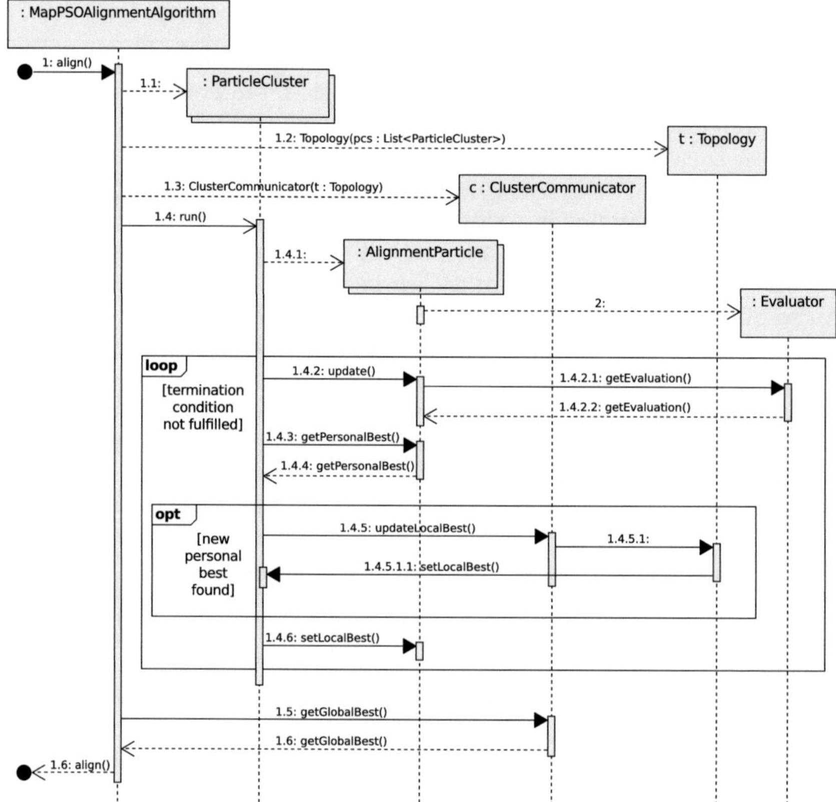

Figure 6.8.: UML sequence diagram of the MapPSO algorithm.

new best alignment to all neighbouring clusters, which consequently update their local best (neighbourhood best) alignment states. Since such updates can come from any neighbouring cluster, the cluster reports the local best state in each iteration to all of its `AlignmentParticle` objects.

6.5. Deployment

There are many ways to make use of an ontology alignment algorithm, ranging from a standalone ontology alignment application to being completely embedded in another semantic application. The two algorithms

Table 6.5.: Implementation classes for building up particle neighbourhoods and communication strategies in MapPSO.

ParticleCluster	
`LocalCluster`	Particle cluster located on the local machine.
`AWSCluster`	Particle cluster located on an Amazon^{TM} EC2 instance (cf. Section 6.5.3).

Topology	
`StarTopology`	All particle clusters are connected to all other particle clusters.
`RingTopology`	Every particle cluster has two neighbours, thus forming a ring of clusters.
`FourClusters`	Particle clusters are grouped into four groups, where all particle clusters in a group are connected as in the `StarTopology`. In each group there are three disjoint particle clusters building the bridges to the three other groups.

ClusterCommunicator	
`SynchronousCluster-Communicator`	In every iteration, particle clusters are waiting for each other before communicating their new best alignment to their neighbours.
`AsynchronousCluster-Communicator`	Particle clusters communicate with their neighbours without waiting for each other (cf. Section 6.5.3).

presented in this book, MapPSO and MapEVO, are deployed in various ways in order to allow for different levels of integration with users or applications.

In order to preserve independence of the concrete alignment algorithm, the various means of deployment have been realised generically in the *KADMOS* alignment infrastructure.

6.5.1. Application Programming Interface

The most low-level way of using any *KADMOS*-based alignment algorithm is to directly import, instantiate, and invoke methods of the *KADMOS* `AlignmentAlgorithm` or `UpdatingAlignmentAlgorithm` interfaces (cf. Figure 6.6). To this end, the corresponding interfaces (*KADMOS*) and classes (MapPSO, MapEVO, or any other *KADMOS*-based algorithm) must be provided in the Java^{TM} classpath or – more conveniently – provided as Apache Maven^{TM} dependency in case the semantic application makes use of Apache Maven^{TM} software management.

There are two restrictions for this kind of algorithm usage. First, since *KADMOS* is a Java^{TM} API, only Java^{TM} applications can directly access the API. Second, since *KADMOS* is based on the OWL API, the semantic application's internal ontology representation must be realised using the OWL API, as well[5].

6.5.2. Web Service

A language independent and decoupled way of using any *KADMOS*-based alignment algorithm is via a Web Service interface. To this end, alignment algorithms can be deployed as SOAP[6] 1.1 [26] Web Service. *KADMOS* provides the functionality to deploy and launch any *KADMOS*-based alignment algorithm as Web Service that is accessible from any remote location.

The advantage of this way of communication with an alignment algorithm is that is independent of the implementation language of the caller due to the open WSDL/SOAP standard. However, a disadvantage is the longer response time, since alignment service and caller are only loosely coupled, *i.e.* input ontologies, as well as resulting alignment are passed in terms of dereferenceable URLs, and each call of the alignment service causes a potential reloading of the ontologies and the alignment[7].

This sort of Web Service deployment has been done in order to participate in the first *SEALS* evaluation campaign[8].

6.5.3. Cloud Infrastructure

Exploiting the inherent parallelisability of population-based optimisation algorithms, the PSO-based alignment algorithm developed in this book has been deployed to a cloud infrastructure [17]. This section gives some background on cloud computing and discusses the challenges and experiences when porting the algorithms on the Amazon Web Services[TM]

---

[5]It is always possible to implement proprietary wrappers, such that OWL API objects can be passed to and from the alignment algorithm, however, this overhead should give reason to think of other ways of integrating the alignment algorithm in the first place.

[6]Simple Object Access Protocol

[7]Surely, various ways of caching and ontology/alignment "diffs" in the case of repeated calls, would be straightforward extensions that diminish the communication bottleneck.

[8]For a full description of the first *SEALS* evaluation campaign, see
`http://www.seals-project.eu/seals-evaluation-campaigns/`
`1st-evaluation-campaigns/ontology-matching`
(accessed January 10, 2012)

(AWS) infrastructure. There are experiments regarding network communication related issues, which is most crucial for the PSO-based alignment approach, due to the social component in the particle swarm that requires more intensive communication between individuals than an Evolutionary Algorithm. For this reason, the following paragraphs focus on the cloud deployment aspects of the MapPSO algorithm.

Background on Cloud Computing

Cloud computing is a new paradigm that has been evolving over the last few years. Most definitions of cloud computing have in common that cloud computing offerings can be categorised using the "Everything-as-a-Service" (XaaS) model. A more detailed view of this model is the "Cloud Computing Stack" [81]. According to the XaaS model the three main service classes are "Software-as-a-Service" (SaaS), "Platform-as-a-Service" (PaaS), and "Infrastructure-as-a-Service" (IaaS). While SaaS offerings usually provide an interface directly to the end user by providing a Web Service interface or a graphical user interface (GUI) the PaaS and IaaS offerings can be used by software architects to build new SaaS services on top of them. PaaS offerings usually provide a platform where the software developer can deploy the new services [81].

The IaaS offerings at the "Basic Infrastructure Services" level, give the developer full control over the servers that are running his software. At this level the user can deploy new machines using Web Service technologies. This offers the power and flexibility to work on a machine level without running one's own data centre and so having convenient access to new resources. Offerings such as the Amazon Elastic Compute Cloud (EC2) give users the opportunity to automatically deploy hundreds of virtual machines within minutes. Thus, it is possible to build highly scalable applications that can scale up and down in short periods. One famous example of a successful cloud offering is Animoto[9]. The Animoto application transforms music files and photos into small video slide shows. After offering their service to users on a social network the demand of virtual machines went from about 40 to 3500 [78]. This scalability was only possible by designing the Animoto software as a distributed algorithm deployed on Amazon EC2.

---

[9]http://animoto.com/
(accessed January 10, 2012)

Another interesting feature of cloud offerings is the typical pay-as-you-go pricing model. This means that users only pay for the resources they are really using. Having such a pricing model it makes no difference if one single server is running for 10 hours or if 10 servers are running for just one hour. The New York Times used this pricing scheme when they built their "TimesMaschine". By using Amazon EC2 they were able to convert their whole archive (4 TB of scanned TIFF files), spanning the years 1851-1922, to Web documents within 24 hours and total costs of US$ 890 [64].

In the context of semantic technologies and the Semantic Web, cloud computing technologies have been successfully applied, mainly for RDF storage [119, 94, 91], querying [91], and materialisation of RDF/OWL knowledge bases [129, 128].

Deployment

The deployment of MapPSO in Amazon EC2 has been realised using a server-worker pattern as depicted in Figure 6.9. Several *workers* evaluate and update several local particles each, while they are managed by a central *server*. Each worker determines the local best alignment among the results of its particles and sends it to the server. The server determines the global best alignment, then broadcasts it and synchronises the workers. Exchange of information between server and workers is realised by the Amazon Simple Queue Service (SQS)[10] or via TCP/IP. The exchange of concrete particle states is realised by the Amazon Simple Storage Service (S3)[11] for reasons of scalability, reliability, and parallel access.

Challenges

Deploying the algorithm on virtual machines connected by a network bears two main challenges. Firstly, the communication latency is much higher when communicating via network than via main memory. Hence finding and broadcasting the global best particle leads to a higher communication overhead, which slows down the algorithm and creates unwarranted costs.

---

[10]`http://aws.amazon.com/sqs/`
(accessed January 10, 2012)
[11]`http://aws.amazon.com/s3/`
(accessed January 10, 2012)

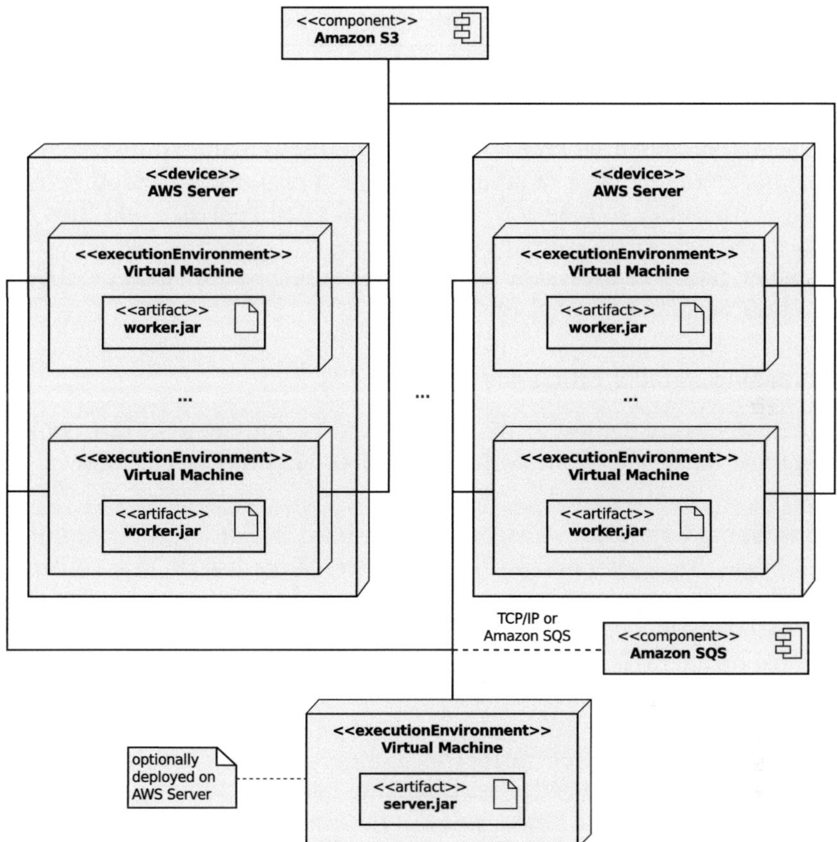

Figure 6.9.: UML deployment diagram for the MapPSO deployment in the Amazon Web Services^{TM} (AWS) cloud infrastructure using the server-worker pattern.

Secondly, the computation times of workers in a particular iteration differ. This difference occurs mainly for two reasons: the unpredictable performance of the virtual environment, and the varying particle sizes. The performance of the virtual machine depends on its mapping to real hardware. For example a "noisy" neighbour, *i.e.* another program sharing the same real hardware via a different virtual machine, can slow down network communication. In turn, a virtual machine can utilise more computing power if there are no or only inactive neighbours. This results in unbalanced performance of the virtual machines. The random initialisation of particles and their different sizes add to this discrepancy in computation time. A small particle needs less computation time than a big particle, therefore a worker with smaller particles will require less runtime per iteration. The random initialisation of particles with different sizes is necessary to search for the optimal alignment size and thus be able to identify *partial* overlaps of ontologies. This computation time discrepancy causes fast workers to idle while waiting for the slower workers and thus decreases parallel efficiency.

Increasing Parallel Efficiency

The challenges of deploying the algorithm to a cloud-based infrastructure identified have been addressed and solutions are proposed as follows.

Addressing Latency. Amazon SQS was used as a means for communication between server and workers. Amazon advertises SQS as reliable, scalable, and simple. However, it turned out that SQS has high latency. For reducing the network latency, direct communication has been implemented using the TCP/IP protocol. Apart from reduced latency, this also resulted in a reduction of the additional communication overhead caused by multiple workers.

The latency is reduced further by only sending particle states when necessary, *i.e.* when a better particle state has been found. To achieve this a worker sends the fitness value of its local best particle to the server and writes the particle state itself into the S3 database. The server then broadcasts the global best fitness and the according database location to all worker instances.

The number of read operations is minimal when using the PSO complete graph topology (cf. Table 6.5), since each particle must know about the global best. In case the global best does not change, no worker has to read a particle state. The (unlikely) worst case in terms of communication latency happens if every worker finds a new best alignment and thus writes its particle state before it has to read a new one found by another worker in the same iteration.

Addressing Runtime Discrepancy. The effect that workers require different runtimes for particle fitness computation can be minimised by *par-*

ticle pooling, i.e. having each worker instance computing multiple particles (particle clusters, cf. Section 6.4, Table 6.5: `AWSCluster`). Using few particles per worker results in high runtime variance between workers caused by different particle sizes and the resulting uneven workload. Using more particles per worker, *i.e.* the workload for each worker being the sum of workloads contributed by each of its particles, averages the runtime because a rather uniform distribution of small and big particles per worker is expected. Using a multi-particle approach also increases parallel efficiency due to the increased overall runtime required to evaluate and update several particles. By increasing the runtime the proportion of time used for communication decreases, which, on the other hand, improves parallel efficiency.

Using *asynchronous* particle updates (cf. Table 6.5: `Asynchronous-ClusterCommunicator`) is another way to compensate the runtime discrepancy. When using synchronous particle updates every worker has to wait for the slowest worker and thus is wasting computation time. In the asynchronous communication mode workers keep on evaluating and updating their particles until they find a particle state that is better than the global best they know about. They send this particle state to the server and continue. The server broadcasts the new particle state if it is better than the global best known by the server. Preventing workers to idle drastically increases parallel efficiency.

Introducing an asynchronous particle update strategy has an effect on the underlying particle swarm metaheuristic. Having not all particles exchanging information at the same time step has a similar effect than changing the social network structure of the PSO from a complete graph topology[12] to a cluster topology[13] [44, Chap. 12] (cf. Table 6.5), which results in fast communication between particles on the same worker compared to the communication between workers themselves. A clustered (*lBest*) topology in general results in slower convergence but better robustness compared to the complete graph (*gBest*) topology [44, Chap. 12].

Furthermore, a loss of information can be observed compared to the synchronous update mechanism. This is due to the fact that particles

---

[12]In the literature the complete graph topology is sometimes (incorrectly) referred to as "star topology".

[13]While in a complete graph topology, every particle shares information with every other particle, the cluster topology allows only groups of particles to communicate with each other, while the groups themselves exchange information only via dedicated particles, which are part of two clusters.

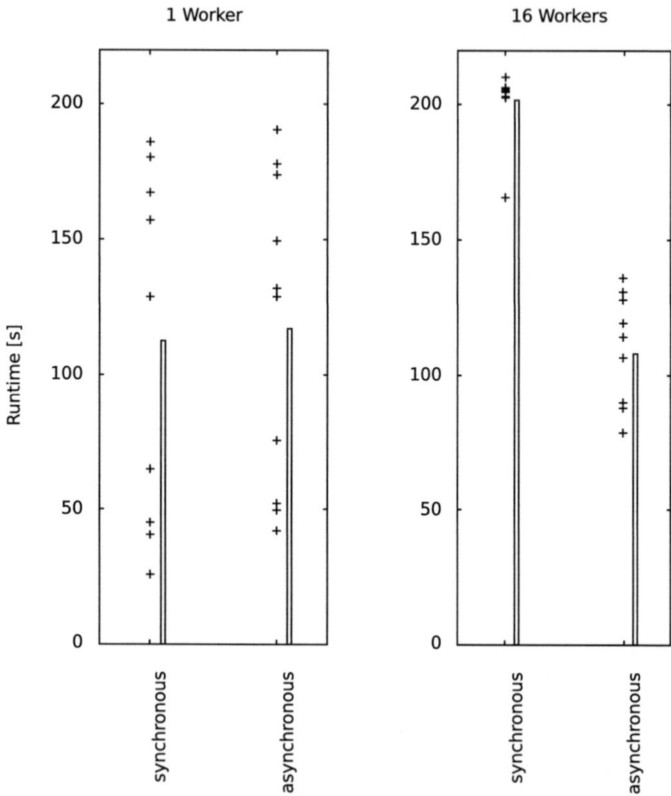

Figure 6.10.: Synchronous vs. asynchronous particle updates using 1 and 16 workers with 1 particle per worker. Bars denote an average of 10 runs (individual runs denoted left of the bars). Depicted is the total runtime for aligning the mouse ontology from the OAEI *anatomy* track (cf. Section 7.2.3) with itself.

compute iterations with possibly outdated information about the global best, which might be available on a worker that is still computing another particle. The problem has been analysed by Lewis *et al.* [84] for parallel multi-objective PSO in heterogeneous and unreliable computing environments. Their investigations revealed that the information loss can be compensated by the increased number of computations that are possible due to the reduced waiting time.

The beneficial effect on the runtime for an alignment is reflected in Figure 6.10, showing runtime behaviour for two medium sized ontologies for both synchronous and asynchronous particle updates. As expected, using an asynchronous communication mode does not have any effect if only a single worker is used. However, for 16 workers that need to communicate their local best results in each iteration, a clear runtime improvement of almost 50 % can be observed.

One further effect resulting from the asynchronous particle updates is that workers hosting mainly small particles can complete an iteration more quickly than those hosting mainly large particles. Therefore small particles tend to compute more iterations and thus influence the swarm stronger than it would be the case in the synchronous mode. This is beneficial to the overall runtime of the algorithm, because the average size of particles is smaller and therefore iterations are faster.

Final Remarks on the Cloud Deployment. The presented cloud deployment of the MapPSO system exploits the straightforward parallelisability of population-based optimisation algorithms. Apart from the technical issues that had to be addressed, cloud deployment bears chances, such as dynamic scalability that now are ready to be exploited by ontology alignment algorithms. Together with the Web Service interface presented in Section 6.5.2, a dynamically scalable alignment cloud service (Alignment-as-a-Service) can be provided, ready to meet requirements arising from the upcoming on-demand, pay-per-use Service Web.

6.5.4. *SEALS* Tool Package

Since 2010, the annual Ontology Alignment Evaluation Initiative (OAEI) evaluation campaign is carried out via the *SEALS* evaluation platform[14] (cf. Section 7.2). From 2011 on, alignment tool developers, who intend to participate in this established evaluation campaign, need to package their tools in a way[15] [89] that it can be automatically deployed and invoked on the *SEALS* platform.

---

[14]`http://about.seals-project.eu/`
(accessed December 19, 2011)

[15]`http://oaei.ontologymatching.org/2011/seals-eval.html`
(accessed March 21, 2012)
and `http://oaei.ontologymatching.org/2011.5/seals-eval.html`
(accessed March 21, 2012)

This packaging has been done for MapPSO and MapEVO, in order to participate in the respective evaluation campaigns (cf. Section 7.2). Technically, the preparation of the *SEALS* tool package involves the implementation of a tool bridge that implements a unified interface common to all ontology alignment tools, and delegates the alignment request to a concrete tool. Besides the tool bridge, a descriptor file must be provided describing the content of the complete tool package. The package itself is bundled as a ZIP file containing the descriptor, the tool bridge implementation, as well as the tool itself including all library dependencies.

Note on the *SEALS* Tool Packaging. Since the *SEALS* tool package has to contain all library dependencies required by the bundled tool, there is the risk of potential conflicts with libraries used within the *SEALS* platform. The *SEALS* platform in the version used for the evaluation campaigns relevant for MapPSO and MapEVO caused such library conflicts. Some of those conflicts could be resolved by removing the respective libraries from the tool package, so that the bundled tool is no longer self-contained, but still executable in the platform. However, other conflicts were nested deep in indirect dependencies and could not be resolved easily, which led to the necessity to deactivate reasoning-based evaluation metrics (cf. Section 4.1).

6.6. Development

All software projects described in this chapter, namely *KADMOS, HARMONIA* Commons, MapPSO, and MapEVO, are publicly available as source code and binaries. The open source development and hosting platform *sourceforge*[16] is used for all software development related issues, such as source code version control (Subversion), issue tracking, and mailing lists.

In particular, the software project *KADMOS* is being developed in the sourceforge project located at

```
https://sourceforge.net/projects/kadmos/
```

while the other software projects *HARMONIA* Commons, MapPSO, and MapEVO are being developed in the sourceforge project located at

---

[16]`http://sourceforge.net/`
(accessed January 11, 2012)

Table 6.6.: Apache Maven^{TM} repositories.

Repositories	
Releases	`http://mavenrepo.fzi.de/semweb4j.org/repo`
Snapshots	`http://mavenrepo.fzi.de/semweb4j.org/snapshots`

Table 6.7.: Apache Maven^{TM} identifiers.

Component	group ID	artifact ID

KADMOS
  API
  Command line
   utilities
  Web utilities
  INRIA API
   wrapper
	`de.fzi.kadmos`	`kadmos`
		`kadmos-api`
		`kadmos-cmdutils`
		
		`kadmos-webutils`
		`kadmos-INRIA-wrapper`

HARMONIA Commons Evaluators	`de.fzi.harmonia.` `commons`	`harmonia-commons` `harmonia-basematchers`

MapPSO
  Core
  AWS Worker
  SEALS Tool Bridge
	`de.fzi.mappso`	`MapPSO`
		`MapPSO-core`
		`MapPSO-awsWorker`
		`MapPSO-seals-toolbridge`

MapEVO
  Core
  AWS Worker
  SEALS Tool Bridge
	`de.fzi.mapevo`	`MapEVO`
		`MapEVO-core`
		`MapEVO-awsWorker`
		`MapEVO-seals-toolbridge`

```
https://sourceforge.net/projects/mappso/
```

All software is being developed using the Java^{TM} (JDK 6) programming language, due to its platform independence and the availability of powerful APIs in particular regarding Web technologies (OWL API, AWS, Web Services, *etc.*).

For smooth project and dependency management, Apache Maven^{TM} is used. Apart from managing dependencies of the projects developed, Maven^{TM} allows for the publication of developed components as artifacts, such that they can be used as dependencies in other projects themselves (cf. Section 6.5.1). To this end, software artifacts are available in remote Maven^{TM} repositories as listed in Table 6.6. The Maven identifiers required by developers that intend to use the components as dependencies are listed in Table 6.7.

On October 5th 2011, MapPSO was included to the Softpedia Mac OS software database, and on March 27th 2012 to Softpedia's database of software programs for the Windows operating system. The promotion via the Softpedia platform includes the Softpedia "100 % Free" award[17], which means that MapPSO was thoroughly tested for spyware, adware, viruses, *etc.*

---

[17]`http://mac.softpedia.com/progClean/MapPSO-Clean-106454.html` (accessed October 25, 2011)
and `http://www.softpedia.com/progClean/MapPSO-Clean-210834.html` (accessed March 27, 2012)

7. Evaluation

The suitability of applying biologically-inspired optimisation techniques
to the ontology alignment problem is empirically evaluated in this chapter. What is presented are studies about how the two approaches introduced in Chapter 5, namely an Evolutionary Algorithm and a Discrete
Particle Swarm Optimisation algorithm meet the expectations expressed
by the motivational conjectures from Section 1.1. All studies are conducted using the prototypical implementations MapEVO and MapPSO
presented in Chapter 6.

Bellahsene *et al.* recognise that "a fundamental requirement for providing universal evaluation of matching and mapping tools is the exis-
tence of benchmarks." [7]. In the context of ontology alignment there is
an established annual campaign called the Ontology Alignment Evaluation Initiative (OAEI) [49], which provides different test cases in various
evaluation tracks. Over the years the OAEI has become the *de facto* evaluation platform for ontology alignment tools. A problem with evaluating
ontology alignment tools in the OAEI context is that the test cases have
a strong focus on comparing tool outputs with provided reference alignments in terms of the standard metrics precision, recall, and F-measure.
Some tracks consider runtime as another criterion for comparison. Until
the most recent campaign in spring 2012, there was no suitable track for
evaluating the scalability of alignment systems, which is not surprising
due to the lack of high quality reference alignments for large ontologies.
Even the *large biomedical ontologies* track in the latest campaign does not
provide a complete reference alignment, but only a partial one. A second problem with OAEI evaluations is that the tools are required to use a
single configuration for all test cases. This requirement is enforced since
the campaign is carried out via the *SEALS* evaluation platform, where all
tools have to be uploaded as a self-contained bundle (cf. Section 6.5.4) using a single (universal) configuration. Following the hypothesis from the
introductory Conjecture 2, different criteria have to be considered in order
to compute alignments for ontologies with different modelling characteristics. So what is essentially evaluated by enforcing the same configura-

tion for ontologies with different modelling characteristics is the ability of an alignment tool to automatically adjust its own configuration.

Due to these restrictions the evaluation presented in this chapter does not solely rely on OAEI data sets. Instead the evaluation is split into several parts in order to back up the hypotheses denoted in the introductory conjectures that biologically-inspired optimisation techniques are a valid approach for solving the ontology alignment problem. After introducing the basic alignment algorithm performance metrics in Section 7.1, the evaluation storyline of this chapter is as follows:

1. The OAEI data sets are used in Section 7.2 in order to demonstrate that the presented approaches are able to provide high quality alignment results for certain test cases. The analysis is further extended by demonstrating that in many cases where the algorithms perform poorly according to traditional precision and recall metrics, the results are not "far" from the reference alignment according to generalised metrics.

2. Since biologically-inspired optimisation techniques are expected to be independent from the objective function to be optimised, a separate study is conducted in Section 7.3 to investigate the correlation between alignment quality and the configuration of the objective function. This in particular explains the performance in terms of alignment quality for the various OAEI tests, depending on the objective function.

3. The insights gained from the previous study suggest that the alignment quality strongly depends on a suitable objective function for the ontology pair at hand. Another experiment is carried out in Section 7.4 to demonstrate the convergence of the algorithms. The convergence in terms of a continuous improvement of the objective function value is compared to the convergence of the F-measure value for intermediate alignments, which demonstrates a clear correlation.

4. The scalability of the approaches is demonstrated in Section 7.5 by exploiting the inherent parallelisability of the algorithms. This experiment is conducted using the MapPSO prototype in a cloud computing infrastructure by aligning two large biomedical ontologies. Since no reference alignment is available for this test data, the con-

vergence of the algorithm is analysed considering objective function value and alignment size for each particle (population member).

The chapter concludes with a discussion in Section 7.6 summarising how the evaluation results correspond to and underpin the conjectures established in Section 1.1.

7.1. Alignment Algorithm Performance Metrics

Typically, alignment quality assessment is done by comparing the alignment with a gold standard (or reference alignment). This comparison is done on the correspondence level by identifying those correspondences contained in the alignment to be evaluated that are also contained in the reference alignment. This allows for the computation of *precision, recall,* and *F-measure* scores, known from information retrieval. Definition 2.7 allows for the application of set intersection on alignments. Let A_1 and A_2 be alignments. Then $A_1 \cap A_2$ is the set of correspondences contained in both A_1 and A_2.

Definition 7.1. Let A be the alignment to be evaluated. Let R be the reference alignment that serves as a gold standard for A. The *precision* of A with respect to R is defined as

$$\mathrm{prec}(A,\, R) = \frac{|A \cap R|}{|A|}$$

The *recall* of A with respect to R is defined as

$$\mathrm{rec}(A,\, R) = \frac{|A \cap R|}{|R|}$$

The *F-measure* is defined as the harmonic mean of precision and recall:

$$F_1(A,\, R) = 2 \cdot \frac{\mathrm{prec}(A,\, R) \cdot \mathrm{rec}(A,\, R)}{\mathrm{prec}(A,\, R) + \mathrm{rec}(A,\, R)}$$

According to this definition the precision of an alignment A with respect to a reference R denotes the fraction of correspondences in A that are correct. The recall of an alignment A with respect to a reference R denotes the fraction of expected correspondences that are contained in A.

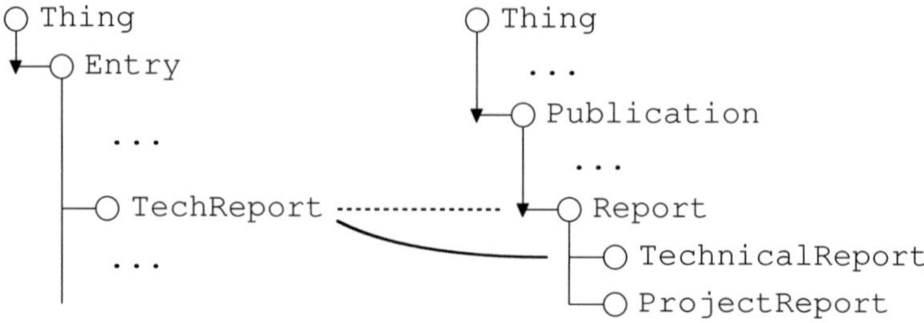

Figure 7.1.: Excerpts from two ontologies with two possible correspondences showing the effect of generalised precision and recall.

The feasibility of these traditional precision and recall metrics in the context of ontology alignment is debatable. The main reason for this scepticism is the fact that these metrics disregard semantics from ontological modelling that might lead to situations where correspondences not contained in the reference alignment are not entirely wrong. For instance, consider the scenario in Figure 7.1, where the correspondence ⟨TechReport, TechnicalReport⟩ (continuous line) would be correct with respect to the reference alignment, while the alignment to be evaluated contains the correspondence ⟨TechReport, Report⟩ (dashed line). Depending on the alignment use case, the correspondence represented by the dashed line is not completely incorrect, since TechnicalReport and Report are in a subsumption relation. For instance, if the alignment is used to incorporate additional data sources for an instance retrieval task, the query results when retrieving all technical reports would be augmented by all instances of Report from the additional ontology. The result would contain all instances of TechnicalReport, thus the result set would be complete. If the end user can cope with some false positives, such as the instances of ProjectReport, the correspondence does no harm.

In order to cope with these kinds of situations, Ehrig and Euzenat introduced the notions of generalised precision and recall [42]. To this end, a *proximity function* is introduced, which replaces the simple correspondence count for the intersection of alignment and reference. Instead the proximity function considers the "closeness" between correspondences from the alignment under evaluation and correspondences from the reference alignment.

In order to be a real relaxation, the proximity function must not return a value lower than the correspondence count for the intersection of alignment and reference. On the other hand, it must not return a value greater than the size of the smaller of alignment and reference. Thus, a proximity function $\omega(A, R)$ of an alignment A and a reference alignment R is bounded as follows:

$$\forall A, R: \quad |A \cap R| \leq \omega(A, R) \leq \min\{|A|, |R|\} \tag{7.1}$$

Definition 7.2. Let A be the alignment to be evaluated. Let R be the reference alignment that serves as a gold standard for A. Let ω be a proximity function. The *generalised precision* of A with respect to R in terms of ω is defined as

$$\text{prec}_\omega(A, R) = \frac{\omega(A, R)}{|A|}$$

The *generalised recall* of A with respect to R in terms of ω is defined as

$$\text{rec}_\omega(A, R) = \frac{\omega(A, R)}{|R|}$$

The *generalised F-measure* is defined in the same way as the classical F-measure from Definition 7.1.

Ehrig and Euzenat proposed three implementations of the proximity function resulting in three generalised precision and recall metrics for ontology alignment. One such implementation is *symmetric precision and recall*, where for each correspondence C in an alignment A the proximity function adds 1, if the correspondence also occurs in the reference alignment R. If R contains a correspondence that is close to C, *i.e.* where one of the corresponding entities is a direct super- or subentity of the one in C, the proximity function adds 0.5. In all other cases the proximity function adds 0. In the example from Figure 7.1 the correspondence ⟨TechReport, Report⟩ would score 0.5 for symmetric precision and recall, instead of 0 for classical precision and recall.

7.2. Ontology Alignment Evaluation Initiative

The *Ontology Alignment Evaluation Initiative* (OAEI) is a coordinated effort to evaluate and compare ontology alignment systems since 2005 [49].

The first OAEI in 2005 was a joint endeavour originating from two predecessor events in 2004: The Information Interpretation and Integration Conference (I^3CON) held at the NIST[1] Performance Metrics for Intelligent Systems (PerMIS) Workshop 2004 in Gaithersburg, MD, and the Ontology Alignment Contest at the 3rd Evaluation of Ontology-based Tools (EON) Workshop held at the International Semantic Web Conference (ISWC) 2004 in Hiroshima.

The MapPSO system has continuously participated in the OAEI campaigns from 2008 till 2011 [15, 18, 12, 13] and in the second *SEALS* evaluation campaign in 2012 (OAEI 2011.5)[2] The MapEVO system participated in the OAEI 2011 [13] and in the second *SEALS* evaluation campaign.

Project *SEALS* (Semantic Evaluation at Large Scale). In 2010 and 2011 some OAEI tracks have become part of the evaluation campaigns of the EU 7th Framework Programme (FP7) project *SEALS* (Semantic Evaluation at Large Scale)[3]. Within the project an evaluation platform and infrastructure is set up to evaluate semantic tools of different categories. The project picks up earlier efforts and experiences from the OAEI and transfers them onto the *SEALS* platform in the context of evaluating systems in the category "Ontology Matching Tools". The *SEALS* project is devoted to run two evaluation campaigns. With respect to "Ontology Matching Tools" the first evaluation campaign coincided with the OAEI 2010. Despite not being an official *SEALS* evaluation campaign, the OAEI 2011 was conducted under *SEALS* conditions using the *SEALS* platform in its development status at that time. Not in the traditional sequence of OAEI events, the second *SEALS* evaluation campaign was carried out the beginning of 2012.

The *SEALS* platform is designed to guarantee comparability of tools and repeatability of evaluations. Regarding tools under evaluation this means that a standalone, self-contained version of each tool must be packaged and uploaded onto the *SEALS* tool repository, including its parameter configuration. In the context of the "Ontology Matching Tools" evaluation, several different alignment scenarios (evaluation data sets) are requested to be processed by the tools using the same set of configura-

---

[1]National Institute of Standards and Technology

[2]`http://oaei.ontologymatching.org/2011.5/`
(accessed May 23, 2012)

[3]`http://about.seals-project.eu/`
(accessed December 19, 2011)

tion parameters. The parameter configurations used for MapEVO and MapPSO for the campaigns carried out in the *SEALS* platform are presented in Appendix B.

7.2.1. Benchmarks Track

The *benchmarks* track has been part of the Ontology Alignment Evaluation Initiative (OAEI) since its foundation in 2005, and is organised by INRIA Grenoble Rhône-Alpes, France. The data sets remained roughly constant apart from minor fixes. Since 2011 the data set is not known prior to the execution of the campaign, since it is systematically generated for each evaluation campaign. However, participants are provided with a similar data set prior to the campaign for testing purposes.

Description

This track aims to identify strengths and weaknesses of alignment systems using a systematically generated data set. To this end, an ontology from the bibliography domain is systematically modified by omitting features (or combinations of features) that can be exploited by alignment systems in order to find correspondences. Each modified version has to be aligned with the original ontology, resulting in a total of 103 alignment tasks in 2011. These alterations are mostly covered in test group 2xx, which constitutes the majority of test cases. Test group 1xx is solely concerned about language conformance with respect to the archaic OWL dialects "Lite" and "DL". In the data set used before 2011, four additional real world ontologies were part of the track (test group 3xx), which were to be aligned with the original ontology.

The systematic modification of the ontology comprises the omission or modification of labels, comments, asserted individuals, properties, or the subsumption hierarchy.

In the OAEI 2010 and 2011 the *benchmarks* track was conducted in the context of the *SEALS* project. Additionally, this track was carried out as part of the second *SEALS* evaluation campaign in spring 2012, which was also named OAEI 2011.5.

Apart from the original data set, which was modified only slightly since 2005, the 2011 campaign contained other data sets that were automatically generated using a particular "seed" ontology [111]. To this end, a data set generator using this "seed" ontology creates a set of test cases similar to

the original data set. The generator allows for a systematic modification
of the "seed" ontology, *i.e.* removal of annotations, removal/adding of
classes of a particular hierarchy level, *etc.*, where the extent of the modification with respect to each feature can be parametrised. In 2011, four
data sets were used, namely the original one (*original*, 33 classes, 24 object properties, 40 data properties, 56 named individuals), the original
one used as seed for automatic benchmark generation (*biblio*, same size
as *original*), the `ekaw` ontology from the *conference* track used as seed
(*ekaw*, 74 classes and 33 object properties), and a finance ontology used
as seed (*finance*, 322 classes, 247 object properties, 64 data properties, and
1113 named individuals).

Results

The MapPSO system participated in the *benchmarks* track since 2008; the
MapEVO system participated since 2011. Table 7.1 shows the results in
terms of precision and recall for MapPSO and MapEVO in the years of
participation. The table shows the aggregated results for the three groups
of tests. Reference alignments are publicly available for the years 2008 till
2010 (before the evaluation was conducted via the *SEALS* platform). Using these reference alignments, the evaluation could be repeated outside
the official OAEI setting for MapEVO, which was still under development in 2010. These locally computed evaluation scores are denoted in
parentheses in Table 7.1. Using the available reference alignments, the
individual results for each test case could be repeated locally in terms of
computing precision, recall, as well as generalised precision and generalised recall. The details are presented in Appendix A. It can be observed
that the results are diverse and, as expected, alignment quality drops with
the absence of information that can be exploited by the alignment system.
Note that by definition the generalised precision and recall scores are in
all cases strictly better than the classical metrics.

A comparison between all participating systems in the OAEI 2011 in the
benchmarks track is shown in Table 7.2. In can be observed that MapEVO
and MapPSO could compute alignments for all data sets in the track, even
for the larger ontologies, where "many of the participants were not able
to process them" [47]. Regarding the alignment quality in terms of F-
measure, MapEVO and MapPSO show results below average compared
to other participants. This can be explained on the one hand by the fact
that due to the generic configuration in 2011, which was used for all OAEI

Table 7.1.: Evaluation results for MapPSO and MapEVO in the OAEI *benchmarks* track since 2008 [28, 45, 46, 47]. (2011 results refer to the generated data set using the *ekaw* seed ontology, 2011.5 results refer to the generated data set using the *biblio* seed ontology.) MapEVO scores for 2010 were computed outside the official OAEI campaign [19], since MapEVO was still under development at that time.

System		MapPSO		MapEVO	
Year	Test group	Precision	Recall	Precision	Recall
2008	1xx	0.92	1.00	–	–
	2xx	0.48	0.53	–	–
	3xx	0.49	0.25	–	–
	H-mean	0.51	0.54	–	–
2009	1xx	1.00	1.00	–	–
	2xx	0.73	0.73	–	–
	3xx	0.54	0.29	–	–
	H-mean	0.63	0.61	–	–
2010	1xx	1.00	1.00	(0.96)	(1.00)
	2xx	0.67	0.59	(0.89)	(0.51)
	3xx	0.72	0.39	(0.77)	(0.44)
	H-mean	0.68	0.60	(0.89)	(0.53)
2011	1xx	0.99	0.92	0.99	1.00
	2xx	0.63	0.62	0.54	0.21
	H-mean	0.64	0.62	0.55	0.22
2011.5	H-mean	0.58	0.12	0.43	0.33

tracks, the chosen objective function was not optimal for the *benchmarks* track. On the other hand, no confidence filtering was done as a postprocessing step, which caused the delivered alignments to still contain correspondences of low confidence. The organisers analysed this in terms of confidence-weighted precision and recall, and report that "these measures provide precision increasing for most of the systems, specially edna, MapEVO and MapPSO (which had possibly many incorrect correspondence with low confidence)" [47].

Regarding the second *SEALS* evaluation campaign, *i.e.* OAEI 2011.5, a similar performance regarding alignment quality was achieved. As in the previous OAEI 2011 campaign, no confidence filtering was applied, which might have improved the results. Additionally, MapPSO was con-

Table 7.2.: F-measure results for all participants in the OAEI 2011 *benchmarks* track [47]. ("X" denotes an execution error, "T" denotes timeout of 2 h for a single test case).

System	original	biblio	ekaw	finance
MapEVO	.41	.37	.33	.20
MapPSO	.50	.48	.63	.14
(AgrMaker)	.88	X	.71	.78
Aroma	.78	.76	.68	.70
CSA	.84	.83	.73	.79
CIDER	.76	.74	.70	.67
CODI	.80	.75	.73	X
edna	.52	.51	.51	.50
LDOA	.47	.46	.52	T
Lily	.76	.77	.70	T
LogMap	.60	.57	.66	X
MaasMatch	.59	.58	.61	.61
MapSSS	.84	X	.78	T
Optima	.64	.65	.56	T
YAM++	.87	.86	.75	T

figured in a way that it disregards the computation of property correspondences, favouring execution in those tracks with larger ontologies, where property correspondences are of no importance. This configuration decision was motivated by the single-configuration policy of the OAEI campaigns executed on the *SEALS* platform (cf. Section 7.6). The omission of property correspondences resulted in a significant decrease of the recall score for MapPSO.

Discussion

Two phenomena can be observed when studying the results from the OAEI *benchmarks* track:

1. On the one hand, for some test cases the algorithms are able to compute alignments of very good quality in terms of precision and recall. On the other hand, there are test cases for which the algorithms perform rather poorly (cf. Appendix A). The alignment algorithms were configured equally for all test cases, and the informa-

tion contained in the ontologies is systematically removed. Thus, the assumption is that the alignment and correspondence evaluation metrics (cf. Chapter 4) implemented in the objective function are sensible for some of the test cases, while they are not sufficient for others. This assumption motivated experiments with alignment and correspondence evaluation metrics, and how they perform for different test cases. The experiments are presented in the following Section 7.3.

2. The performance of the alignment algorithms is strictly better with respect to symmetric precision and recall than for the classical metrics. In their official OAEI result analysis for 2009 the organisers report that "[...] MapPSO has significantly better symmetric precision and recall than classical precision and recall, to the point that it is at the level of the best systems. This may be due the kind of algorithm which is used, that misses the target, but not by far" [45]. Similarly, for the OAEI 2011 the organisers report that with respect to generalised evaluation metrics "[...] the exception is MapEVO, which has a considerable improvement in precision. This could be explained by the fact this system misses the target, by not that far (the false negative correspondences found by the matcher are close to the correspondences in the reference alignment) so the gain provided by the relaxed [*generalised*, author's remark] measures has a considerable impact for this system. This may also be explained by the global optimization of the system which tends to be globally roughly correct as opposed to locally strictly correct as measured by precision and recall" [47]. These observations match the general behaviour of optimisation metaheuristics that they perform a coarse grained search, and an additional local search step is required to fine-tune the results [106, Sect. 6.3].

Criticism. The OAEI *benchmarks* track has become a *de facto* standard for evaluating ontology alignment systems and comparing new systems with the state-of-the-art. However, the data sets should be used with care when reporting about the performance of alignment systems. This holds in particular for the provided reference alignments, as they are systematically generated along with the individual test ontologies. This means that for each alteration of the original ontology, the reference alignment is altered accordingly. It requires special care when this is done automatically, since removing features from the ontology can lead to situations, where there is no or not enough remaining evidence that certain entities participate in a particular correspondence. Such correspondences must not be included in a reference alignment, which is reflecting the gold standard for the particular alignment test case. Unfortunately, such correspondences can be found in the reference alignments provided with the *benchmarks* track. An example from the OAEI 2011 version, where such a situation occurs is test case #201 [13]. The correspondences ⟨Conference_Trip, GBCFRTQEDNXEZMVRUWLFXTDFKC⟩ as well as ⟨Conference_Banquet, KKRDJIPEEQFBQKOWPOPJWENCPL⟩ are part of the reference alignment and both are assigned a confidence value of 100 %. Even though there is evidence – drawn from the position in the subsumption hierarchies – that the two entities from the first ontology correspond to the two entities from the second ontology, it is unclear which corresponds to which. This precise assignment cannot even be done by a human being closely looking at the ontologies, since all features that support any of the two correspondences have been removed in test case #201. Since in this example both correspondences cannot be information theoretically justified, they should not be part of the reference alignment. An alternative remedy would be to reduce the confidence value in the reference alignment for both correspondences to 50 %, since it is still up to the alignment system to *guess* an assignment, however, the chance for a correct guess is at most 50 % in this case.

As a conclusion, the OAEI *benchmarks* track should only be used bearing in mind the fact that in many test cases, precision and recall of 100 % cannot be achieved reliably. Furthermore, it is not documented what the highest achievable precision and recall scores are for each test case, so eventually the data sets can only be used to *compare* systems with respect to precision and recall metrics.

7.2.2. Directory Track

The *directory* track, organised by the University of Trento, was part of the Ontology Alignment Evaluation Initiative (OAEI) from 2005 till 2010. In 2005 a different data set was used than in subsequent years. This 2005 data set "allowed only the estimation of recall" [46, Section 6]. For that reason, and since the recall scores reported for 2005 are below 35 % for all participants [51], that year is spared out for comparative analysis here.

Table 7.3.: Evaluation results for the MapPSO system in the OAEI *directory* track 2008 and 2010 [28, 46].

	Precision	F-measure	Recall
MapPSO (2008)	0.57	0.40	0.31
MapPSO (2010)	0.61	0.60	0.58

Description

The track aims to reflect the real world use case of aligning large Web directories [46, Section 6]. To this end, the data used in the *directory* track is extracted from the Web directories of Google[TM4], Yahoo!®[5], and LookSmart®[6] As Web directories represent simple taxonomies, the ontologies are of low expressiveness mainly consisting of a classification (subsumption) hierarchy. The prepared data set consists of 4639 individual alignment tasks each one being a pair of ontologies from the Web directories. For each task there is a manually created reference alignment. The submitted alignments were compared with these reference alignments, and precision, recall, and F-measure scores were reported.

Results

The MapPSO system participated in the *directory* track in the years 2008 and 2010. Table 7.3 shows the results in terms of precision, recall, and F-measure for the MapPSO system in the years of participation.

The figures show an improvement in precision from 57 % to 61 %, and in recall from 31 % to 58 %. In particular the significant increase in recall caused the F-measure score to increase by one third.

With the data set staying the same since 2006 it was possible to compare the performance of alignment systems over time, as well as the best

---

[4]`http://www.google.com/dirhp`
unavailable since Google[TM] shut down the service in July 2011.

[5]`http://dir.yahoo.com/`
(accessed December 13, 2011)

[6]`http://www.looksmart.com/r?country=uk`
The Web resource is no longer available since March
2006; see `http://www.searchenginejournal.com/`
`looksmart-closes-zeal-concentrates-on-furlnet/3161/`
(accessed December 13, 2011).

performing systems from each year. Figure 7.2 shows the performance of the three best participating systems for each year in terms of precision, recall, and F-measure.

As can be seen from the plots, best precision scores achieved were roughly constant around 60 % from 2007 on. The challenge of the track is to achieve high recall scores. The plots show that those best performing systems of the five years on average achieve about 50 % recall. An exceptional year was 2007, where two of the three best performing systems identified more than 70 % of the expected correspondences (recall).

MapPSO with its results from 2010 is ranked second with respect to F-measure in the that year[7]. Compared to the best performing systems in the history of the track, MapPSO ranks third with respect to F-measure. (The ASMOV system achieved a slightly higher F-measure score of 63 % twice in the years 2009 and 2010.) Regarding recall, MapPSO is ranked fifth (again double counting ASMOV from the years 2009 and 2010), and regarding precision, MapPSO is ranked second. It shall be noted that the difference between the top performing systems here is in the range of a few percent regarding precision, recall, and F-measure.

In addition to the "small" tasks provided from 2006 to 2010, there was a "single" task provided in 2010, where the scalability of alignment systems should be evaluated. MapPSO results were submitted for this task, however, the organisers reported that "the task was cancelled due to lack of resources needed to cross check the reference alignments" [46, Section 6].

Discussion

The results of MapPSO in the OAEI *directory* track have shown that the system is competitive in real world alignment scenarios from the domain of Web directories. Since in the history of the track only "small" tasks were evaluated, this statement refers solely to alignment quality and not to scalability. From the best performing alignment systems in the five year history of the track MapPSO was able to achieve the third highest F-measure score from all participating systems. The fact that the performance of different systems varies only in the range of a few percent gives reason to assume that fine-tuning similarity metrics is an important factor. In particular regarding MapPSO there was no linguistic background knowledge exploited when computing the results. However, linguistic

---

[7]It has to be mentioned that in the year 2010 only three systems participated in the *directory* track.

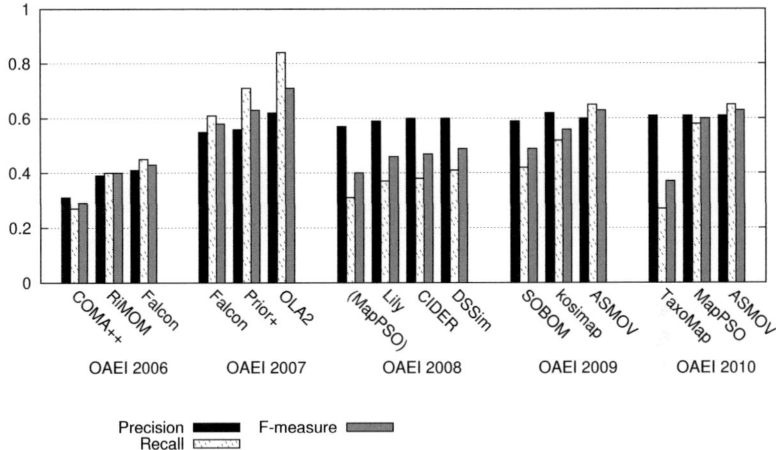

Figure 7.2.: The top 3 performing ontology alignment systems in the OAEI *directory* track for each campaign in the years 2006 till 2010 [46, Section 6]. (Based on `http://oaei.ontologymatching.org/2010/results/directory/files/comparison.png`, accessed December 14, 2011.)

knowledge could play an important role when aligning Web directories, due to the possibly hidden semantics in class labels. For instance in order to correctly identify correspondences for a class labelled "Breweries & Brands", linguistic analysis of the label would be required that goes beyond the purely lexical methods that were used in the configuration of MapPSO for this track.

7.2.3. Anatomy Track

The *anatomy* track, organised by the University of Mannheim, is part of the Ontology Alignment Evaluation Initiative (OAEI) since 2005. Since 2007 the ontologies used in the data set remained the same, however, several improvements have been made since then for both the ontologies and the (manually created) reference alignment.

Description

The track aims to reflect the real world use case of aligning large ontologies from the biomedical domain. The data set used in the *anatomy* track consists of the two ontologies "Adult Mouse Anatomy" and an excerpt of the National Cancer Institute (NCI) thesaurus[8], as well as a manually created reference alignment [21]. In 2011 the "Adult Mouse Anatomy" contained 2744 classes, the NCI excerpt 3304 classes and the reference alignment 1516 correspondences. In this respect, the *anatomy* track is the largest data set in the OAEI with a high quality reference alignment available. According to the track organisers, the data set contains about 60 % of trivial correspondences in the sense that they can be identified by basic string comparison techniques.

In the OAEI 2010 and 2011 the *anatomy* track was conducted in the context of the *SEALS* project. Additionally, this track was carried out as part of the second *SEALS* evaluation campaign in spring 2012.

The *anatomy* track had four participation modalities until 2012 that allowed systems to modify their configuration parameters in order to maximise precision, recall, F-measure, and to additionally consume a partial initial alignment. Since the *SEALS* platform does not allow for adjustment of configuration parameters, the modalities to maximise precision and recall, respectively, were discarded in the OAEI 2011 and the second *SEALS* evaluation campaign in spring 2012.

Results

The results for each participant in the OAEI 2011 *anatomy* track are listed in Table 7.4. Due to the size of the ontologies, 5 out of 15 systems were not able to compute a result in less than 24 h. The organisers report that "the two systems MapPSO and MapEVO can cope with ontologies that contain more than 1000 concepts, but have problems with finding correct correspondences. Both systems generate comprehensive alignments, however, MapPSO finds only one correct correspondence and MapEVO finds none. This can be related to the way labels are encoded in the ontologies. The ontologies from the anatomy track differ from the ontologies of the benchmark and conference track in this respect" [47]. No alignment quality results of MapEVO and MapPSO were reported by the organisers

---

[8] http://ncit.nci.nih.gov/
(accessed December 15, 2011)

Table 7.4.: Alignment size, precision, recall, and F-measure results for all participants in the OAEI 2011 *anatomy* track [47]. ("X" denotes an execution error, "T" denotes timeout of 24 h).

System	Size	Precision	F-measure	Recall
MapEVO	1079	.00	.00	.00
MapPSO	2730	.00	.00	.00
(AgrMaker)	1436	.94	.92	.89
Aroma	1279	.74	.68	.63
CSA	2472	.47	.57	.76
CIDER	T	T	T	T
CODI	1298	.97	.89	.83
edna	934	1.0	.77	.62
LDOA	T	T	T	T
Lily	1368	.81	.77	.73
LogMap	1355	.95	.89	.85
MaasMatch	1079	1.0	.45	.29
MapSSS	X	X	X	X
Optima	X	X	X	X
YAM++	X	X	X	X

for the OAEI 2011.5 campaign[9].

Outside the official OAEI setting, experiments with MapEVO were conducted using a specific configuration and the *anatomy* track ontologies from the year 2010. Thereby, a larger number of iterations was used and the objective function was adjusted to consider the label encoding used in this track. The results as shown in Table 7.5 demonstrate a significant improvement with respect to the 2010 reference alignment, in particular in terms of precision.

Discussion

The analysis of the organisers reflects the generic configuration of the two systems MapEVO and MapPSO used in the OAEI 2011 campaign. The objective function is statically configured and no adjustments are done

---

[9]`http://web.informatik.uni-mannheim.de/oaei/anatomy11.5/`
`results.html`
(accessed May 23, 2012)

Table 7.5.: Evaluation results of the MapEVO system for the OAEI *anatomy* ontologies 2010. (No official OAEI participation.)

	Precision	F-measure	Recall
MapEVO (2010)	0.82	0.56	0.43

automatically in order to meet track specific requirements. In the case of the *anatomy* track, for instance, the IRI fragments of entities interpreted by the h_{lexIDSim} and $h_{\text{textIDSim}}$ are internal identifiers that cannot be used for evaluating correspondences. The generic configuration used, however, treats them as significant, which perturbs the overall similarity aggregation and thus prevents useful handling of correspondence confidences. The similarity, however, is beneficial in other OAEI tracks, which is the reason why it is incorporated into the objective function used.

In contrast, for instance, the AgreementMaker system follows the approach to use a specific setting tailored for each of the three OAEI track, which is automatically selected during execution—a behaviour which is forbidden for OAEI participation. (See note in Section 7.6.)

In a separate experiment with an adjusted objective function, MapEVO was able to demonstrate a significant improvement. In particular the precision scores above 80 % are competitive. Regarding the recall score of below 50 %, it has to be mentioned that no external background knowledge base was used for similarity computation and correspondence evaluation. However, in order to obtain competitive recall results in the *anatomy* track most systems utilise the UMLS metathesaurus [22]. According to the organisers, the reference alignment contains about 60 % of "trivial" correspondences [48] that do not require external resources in order to be identified. This naturally limits the recall that can be achieved by alignment systems that do not exploit such external knowledge, such as the UMLS metathesaurus. This puts the recall score of MapEVO reported in Table 7.5 into a different perspective.

7.2.4. Conference Track

The *conference* track, organised by the University of Economics, Prague, Czech Republic, was part of the Ontology Alignment Evaluation Initiative (OAEI) since 2006. From 2006 till 2008, the *conference* track was organised as "consensus workshop", where references alignments were not available prior to execution of the campaign. Instead, generated alignments were evaluated manually, and put up for discussion in the Ontology Matching workshop. Since 2009, reference alignments are available and used for evaluation, similarly to the other OAEI tracks[10]. In each year additional ontologies were added to the data set.

Description

The track aims to reflect the real-world use case of aligning ontologies from the domain of conference organisation. Compared to other OAEI tracks, ontologies in the *conference* track are more expressive in terms of the description logic fragments they cover. In 2011 the data set consisted of 16 ontologies of various size and expressiveness[11]. The task is to find correspondences between all ontologies, which participating tools typically handle by computing alignments between all pairs of ontologies.

In the OAEI 2010 and 2011 the *conference* track was conducted in the context of the *SEALS* project. Additionally, this track was carried out as part of the second *SEALS* evaluation campaign in spring 2012.

Results

The results for each participant in the OAEI 2011 *conference* track are listed in Table 7.6. The performance of both MapPSO and MapEVO is clearly below average in this track. Due to the deactivation of property correspondence computation (see also Section 7.2.1) the recall score for MapPSO decreased to 0.05 in the OAEI 2011.5 campaign. For MapEVO the result (in terms of F-measure) remained roughly similar compared to 2011.

Discussion

The poor performance of MapPSO and MapEVO in this track may have several reasons. Firstly, the single configuration used for all tracks in the OAEI 2011 and 2011.5 participation was not taking into account complex ontology language features as found in the *conference* track. For instance, ontologies in this track frequently contain complex class descriptions in

---

[10]`http://nb.vse.cz/~svatek/ontofarm.html`
(accessed December 19, 2011)
[11]`http://nb.vse.cz/~svabo/oaei2011/`
(accessed December 19, 2011)

Table 7.6.: Precision, recall, and F-measure results for all participants in the OAEI 2011 *conference* track [47].

System	Precision	F-measure	Recall
MapEVO	.15	.04	.02
MapPSO	.21	.23	.25
(AgrMaker)	.65	.62	.59
Aroma	.35	.4	.46
CSA	.5	.55	.6
CIDER	.64	.53	.45
CODI	.74	.64	.57
LDOA	.1	.17	.56
Lily	.36	.41	.47
LogMap	.84	.63	.5
MaasMatch	.83	.56	.42
MapSSS	.55	.51	.47
Optima	.25	.35	.57
YAM++	.78	.65	.56

subclass axioms, which are not considered and interpreted by the correspondence and alignment evaluators used in the objective function. Thus the objective function used could not grasp the characteristics of the ontologies used in the *conference* track. Secondly, the entity labels used in this track demonstrate several linguistic facets that are not taken care of by any of the used correspondence evaluators. In particular normalisation and the analysis of compound words and phrases seems to be beneficial for this track, but is not done by any of the evaluators used in MapPSO and MapEVO.

7.2.5. Large Biomedical Ontologies Track

The *large biomedical ontologies* track, organised by the University of Oxford, United Kingdom, was introduced as part of the Ontology Alignment Initiative (OAEI) in the context of the second *SEALS* evaluation campaign in spring 2012. The track was set up in the context of the LogMap project[12] that has a focus on providing logically correct mappings between large

---

[12]http://www.cs.ox.ac.uk/isg/projects/LogMap/
(accessed April 2, 2012)

biomedical data sets, such as SNOMED Clinical Terms (SNOMED CT)[13], the Foundational Model of Anatomy (FMA)[14], and the National Cancer Institute Thesaurus (NCI)[15].

Description

From the real-world problem of obtaining correspondences between concepts of the large biomedical ontologies, an OAEI track was set up as a challenge for ontology alignment tools. The track consists of only a single pair of ontologies, namely FMA and NCI. However, each of the two ontologies has been reduced in its size in order to have three tasks in this OAEI track of differently sized ontologies[16]:

1. Small ontology fragments of overlapping parts. The FMA module contains 3,696 concepts, which is 5 % of the complete FMA. The NCI module contains 6,488 concepts, which is 10 % of the complete NCI.

2. Big ontology fragments of overlapping parts. The FMA module contains 28,861 concepts, which is 37 % of the complete FMA, while the NCI module contains 25,591 concepts, which is 38 % of the complete NCI.

3. Complete ontologies with the FMA containing 78,989 concepts and the NCI containing 66,724 concepts.

There are two reference alignments provided that were extracted using the UMLS metathesaurus [22], where one reference is the unmodified UMLS mapping (3,024 correspondences), and the other reference is repaired in order to avoid inconsistencies (2,898 correspondences). Both references are not meant to be complete, since a manually validated, complete alignment between FMA and NCI does not exist. It is one of the goals of the track organisers to use the OAEI challenge to extend the UMLS baseline mappings and provide a "silver standard" for advancing in this biomedical use case scenario.

---

[13]http://www.ihtsdo.org/snomed-ct/
(accessed April 2, 2012)
[14]http://sig.biostr.washington.edu/projects/fm/
(accessed April 2, 2012)
[15]http://ncit.nci.nih.gov/
(accessed April 2, 2012)
[16]http://www.cs.ox.ac.uk/isg/projects/SEALS/oaei/
(accessed April 2, 2012)

Results

The result for the *large biomedical ontologies* track were provided by the organisers only for new or modified systems compared to the OAEI 2011 campaign, and for those participants that could process the ontologies from the OAEI 2011 *anatomy* track. Thus results were reported for 10 out of 19 participating systems that were able to process the large biomedical ontologies. All of these 10 systems could cope with the small overlapping ontology fragments, while only 9 system could cope with the big overlapping fragments and the complete ontologies. MapEVO and MapPSO were able to process all inputs. The participating systems were "executed [...] in two different settings: (1) a standard laptop with 2 cores and 4 GB of RAM, and (2) a high performance server with 16 CPUs and 10 GB [*of RAM*, author's remark]"[17]. For the small overlapping ontology fragments task, all systems except one were able to compute the alignment in the conventional "laptop" setting. Out of the 10 participating systems, for the big overlapping ontology fragments task only 8 systems, and for the whole ontologies only 7 systems could be run in the "laptop" setting. The presented approaches MapEVO and MapPSO, however, were able to complete the tasks in this restricted setting.

Table 7.7 shows the results of all examined participants for the small overlapping ontology fragments task regarding the unmodified UMLS reference. The results of MapEVO and MapPSO regarding the precision and recall scores in this task are clearly below average. While in the small overlapping fragments task scores of less than 5 % were achieved, the results are 0 for the big overlapping fragments and whole ontologies tasks. These low scores were expected, in particular for the two larger test cases, since a configuration was used that did not unveil the full potential of the presented algorithms. To be specific, a low population size was used due to the restrictions of the *SEALS* evaluation platform (cf. Section 7.6). Moreover, a small number of iterations was used in order to meet the time limits imposed by the evaluation campaign setting.

Discussion

This track demonstrates the ability of the MapEVO and MapPSO prototypes to process large ontologies. Only 10 out of the 19 systems partic-

---

[17]http://www.cs.ox.ac.uk/isg/projects/SEALS/oaei/ results2011.5. html
(accessed May 23, 2012)

Table 7.7.: Alignment size, precision, recall, and F-measure results for all participants in the OAEI 2011.5 *large biomedical ontologies* track (small overlapping ontology fragments) drawn from `http://www.cs.ox.ac.uk/isg/projects/SEALS/oaei/results2011.5.html`, accessed May 23, 2012. The scores refer to the unmodified UMLS reference.

System	Size	Precision	F-measure	Recall
MapEVO	633	.003	.002	.001
MapPSO	3654	.022	.024	.027
Aroma	2575	.824	.758	.702
CSA	3607	.528	.574	.629
GOMMA-bk	2878	.957	.933	.910
GOMMA-nobk	2628	.973	.905	.846
LogMap	2739	.952	.905	.863
LogMapLt	2483	.969	.874	.796
MaasMatch	3696	.597	.657	.730
MapSSS	1483	.860	.566	.422

ipating in the OAEI 2011.5 were able to process those ontologies, and, regarding the whole ontologies task, only 7 out of those 10 systems could perform on restricted hardware resources. MapPSO and MapEVO are able to be executed on those limited resources. Additionally, MapPSO and MapEVO are capable of utilising multiple compute nodes due to their parallel algorithm architecture, which, however, could not be fully utilised in this evaluation campaign setting. On the other hand, the results demonstrate that the latest versions of some competing ontology alignment systems are indeed also able to process large ontologies provided that sufficient hardware resources are available.

Regarding the alignment quality results, the low scores mainly go back to the trade-off parameter configuration used in order to obey the single-configuration policy of the OAEI. On the other hand, the small population and the low number of iterations used in the setting lead to significantly worse scores than if those were set sufficiently large. A sufficient configuration is not a problem in distributed execution environments (cf. Section 7.5), but cannot be realised in the resource and time restricted setting of the evaluation campaign.

7.3. Effectiveness of Evaluation Metrics

The evaluation results from the previous Section 7.2 show a diverse performance of the introduced prototypes MapEVO and MapPSO regarding alignment quality. While for some test cases, such as *benchmarks* 103 the results with respect to precision and recall are very good, there are other cases, *e.g. anatomy*, where the results are rather poor. An interesting observation in the context of the *anatomy* track was the fact that the same algorithm (MapEVO) revealed significantly better results when executed locally on the same data set with an adapted objective function. This section investigates the correlation between algorithm performance in several test cases and the performance of the alignment evaluation metrics introduced in Chapter 4. To this end the *discriminatory* behaviour of the evaluation metrics is analysed. "Discriminatory" in this case means the property of an evaluation metric to provide a high score for a very good alignment, and a low score for a very bad alignment. This behaviour is indispensable for the optimisation algorithms to distinguish between good and bad solutions and thus converge towards the optimum.

In order to measure the discriminatory behaviour of the evaluation metrics their evaluation scores are computed for some of the OAEI test cases' reference alignments, which are supposed to represent the optimal alignment for those cases. Additionally, the evaluation metrics are applied to five randomly generated alignments for the ontologies of the same test cases, with their average reported. It is assumed that a random alignment is a rather bad solution candidate for the alignment problem, so it is expected that the average of the evaluations of those random alignments is relatively low compared to the evaluation of the reference alignment. In order to have a single evaluation score for correspondence level evaluators, the evaluation scores of all correspondences in the alignment are aggregated by computing the arithmetic mean[18]. Note that this aggregation hides the impact of correspondence level evaluators when their evaluation scores are used as *support heuristics* in the optimisation algorithms. Hence, a correspondence level evaluator that has poor discriminatory behaviour on the alignment level as presented here, may still be valuable in terms of distinguishing good from bad correspondences.

---

[18]This is the same computation as done by the correspondence contribution alignment evaluator (cf. Equation (4.43))

7.3.1. Evaluation Metrics in the OAEI *benchmarks* Track

The systematic design of the test suite in the OAEI *benchmarks* track revealed on the one hand good performance for MapEVO and MapPSO, for instance in test case 103, on the other hand bad performance, for instance in test case 252. Figure 7.3 shows the evaluation scores for local correspondence evaluators, contextual correspondence evaluators, and global alignment evaluators, respectively, for the *benchmarks* test case 103.

In particular the local correspondence evaluators show clear discriminatory behaviour, which is less significant for the contextual correspondence evaluators. Here it should be mentioned that the $h^A_{\mathrm{hierarchyProp}}$ metric depends on the evaluations of "neighbouring" correspondences (cf. Equation (4.21)), which is always 1.0 in the reference alignment, and a uniform random value in the random alignments. In the light of this analysis, the highly discriminatory behaviour is still representative, since the low value for the random baseline alignment shows that there are only few neighbouring correspondences at all. In a real optimisation run these neighbouring evaluation scores depend on the evaluation computations from previous iterations, as it is the case for many contextual correspondence evaluators. For the global metrics on the alignment level, the H_{consist} metric is a discrete one that scores either 0 or 1. The metric seems to be valuable since all random alignments induce an inconsistency. Consequently the $H_{\mathrm{coherence}}$ metric scores 0, since due to the inconsistency, all classes have become unsatisfiable. The contextual $h^A_{\mathrm{explanation}}$ is inapplicable for the reference alignment, since there cannot be an explanation if there are no unsatisfiable classes. The same metric is also inapplicable for the random alignments, since they are all inconsistent, and computing explanations cannot be done on an inconsistent ontology.

The results produced by the same evaluators for the OAEI *benchmarks* test case 252, where the overall result of MapEVO and MapPSO was not satisfactory (cf. Section 7.2.1 and Appendix A) are shown in Figure 7.4.

Compared to the results for test case 103, it can be observed that the local correspondence evaluation metrics do not provide high scores for the reference alignment, and consequently show no discriminatory behaviour. The results for the contextual correspondence evaluators and the global alignment evaluators are similar to those for test case 103. The availability of highly discriminatory local evaluation scores seems to be an important factor to achieve good overall alignment performance. An explanation for this observation is that some contextual correspondence

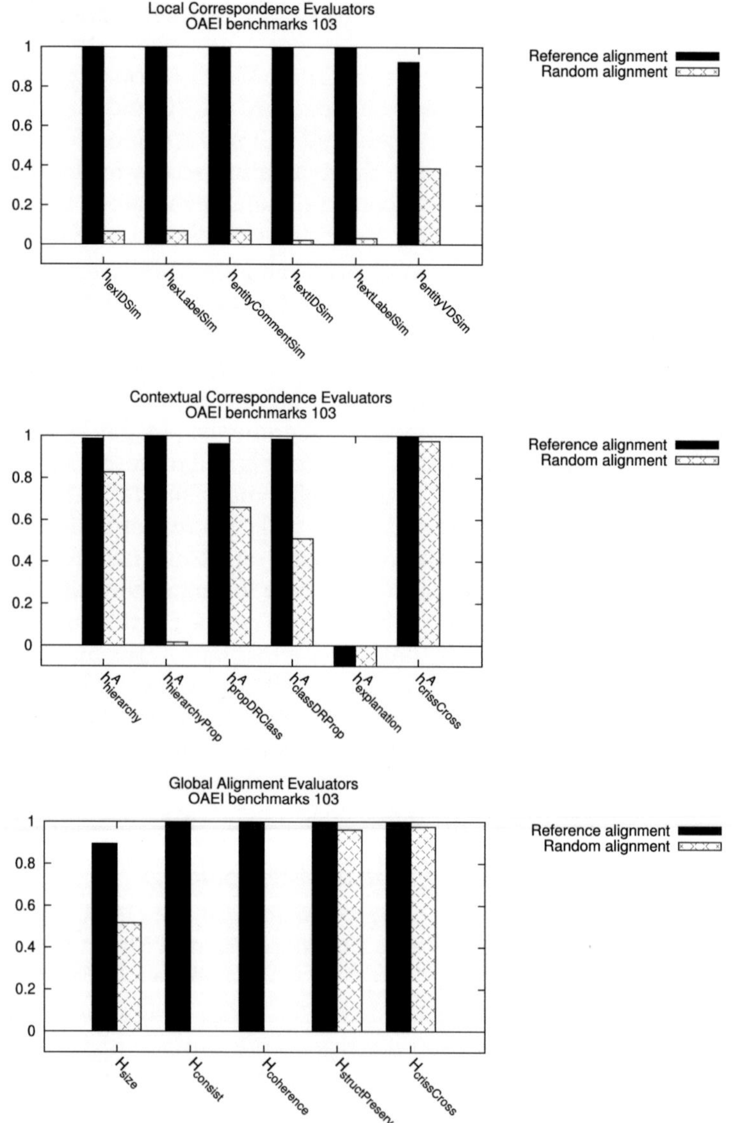

Figure 7.3.: Evaluation scores for different evaluator types for reference and random alignments of test case 103 from the OAEI *benchmarks* track. Inapplicability of an evaluator is indicated by a negative score.

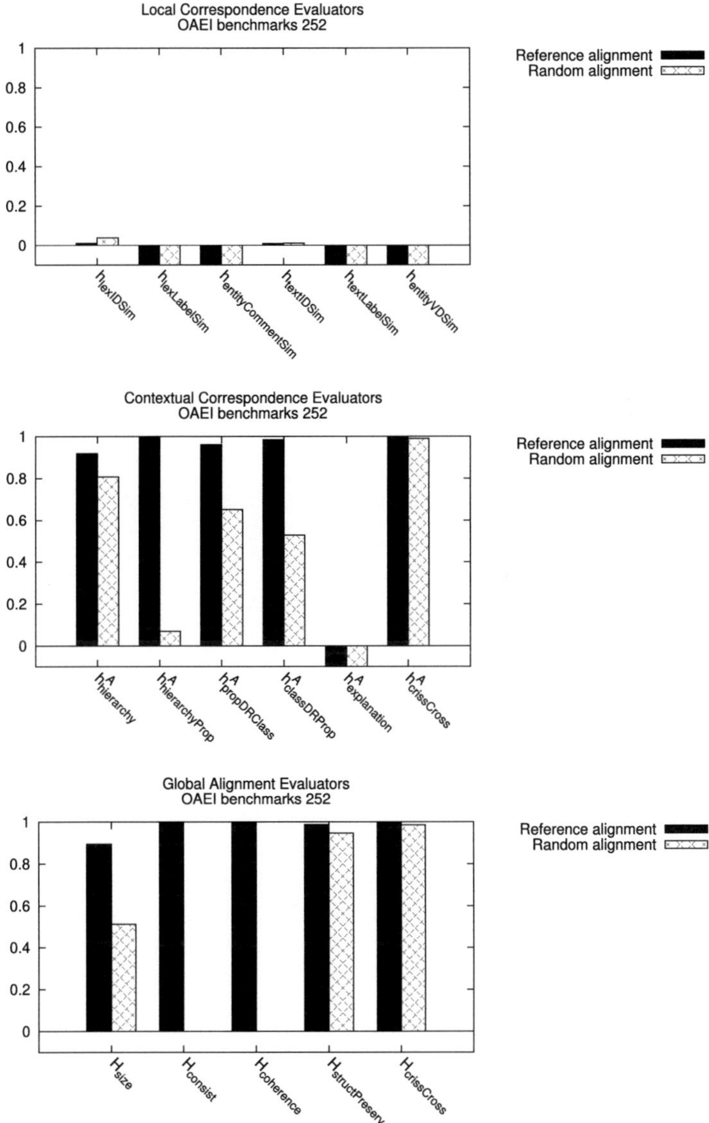

Figure 7.4.: Evaluation scores for different evaluator types for reference and random alignments of test case 252 from the OAEI *benchmarks* track. Inapplicability of an evaluator is indicated by a negative score.

evaluators ($h^A_{\text{hierarchyProp}}$, $h^A_{\text{propDRClass}}$, $h^A_{\text{classDRProp}}$) are propagating evaluation scores from other correspondences to the one under evaluation. So good correspondences should have a high evaluation score in order to influence other correspondence scores accordingly. Local correspondence evaluation metrics are *bootstrapping* the evaluation scores, which can in later iterations be propagated by contextual evaluators. The absence of discriminative local correspondence evaluation metrics hampers the effect of contextual evaluators as well.

7.3.2. Evaluation Metrics in other OAEI Tracks

The discriminatory behaviour of evaluation metrics is studied for two ontologies from the OAEI *conference* track, `ekaw.owl` and `iasted.owl` (Figure 7.5), as well as for the *anatomy* track ontologies (Figure 7.6), and the small sized versions of the ontologies from the *large biomedical ontologies* track (Figure 7.7).

It can be observed that the performance and discriminatory behaviour of evaluation metrics are diverse. This confirms the assumption that ontologies of varying characteristics require different alignment evaluation metrics in order to distinguish good from bad solutions, thus underpinning the introductory Conjecture 2. In the concrete scenarios this becomes most evident for the h_{lexIDSim} and $h_{\text{lexLabelSim}}$ ($h_{\text{textIDSim}}$ and $h_{\text{textLabelSim}}$, analogously) evaluation metrics. While there is significant information in the entity identifiers in the *conference* ontologies, the identifiers are meaningless in case of the *anatomy* ontologies. The opposite holds for the entity labels. Recalling the *anatomy* results from Section 7.2.3 the objective function used in the official campaign did not consider entity labels, but only entity identifiers, which explains the poor alignment quality achieved in the OAEI 2011.

Another observation is the discriminative behaviour of the $h^A_{\text{hierarchy}}$ evaluator for the *conference* track ontologies. Here, the random alignment achieves a better score than the alignment. This most likely goes back to the fact discussed earlier in Section 7.2.4 that the *conference* ontologies contain complex class descriptions in subclass axioms, which are not interpreted by the current implementations of the $h^A_{\text{hierarchy}}$ evaluator. This often leads to situations, where existing correspondences of super- or subclasses are not recognised as such by the evaluator.

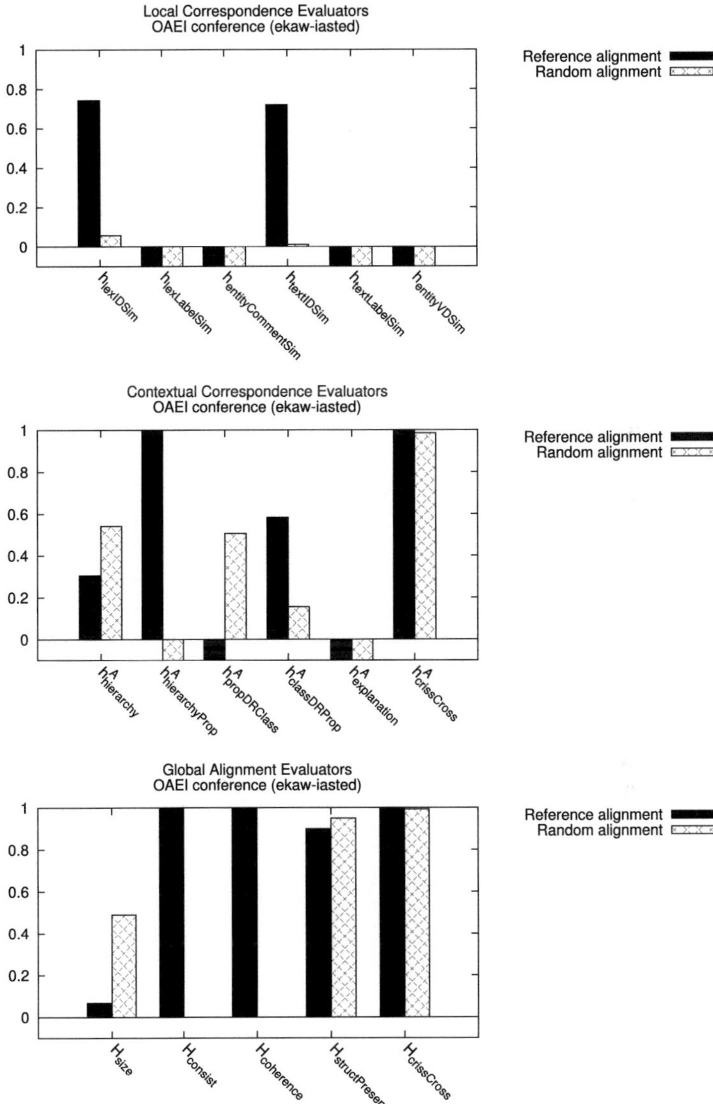

Figure 7.5.: Evaluation scores for different evaluator types for reference and random alignments of two ontologies from the OAEI *conference* track (`ekaw.owl` and `iasted.owl`). Inapplicability of an evaluator is indicated by a negative score.

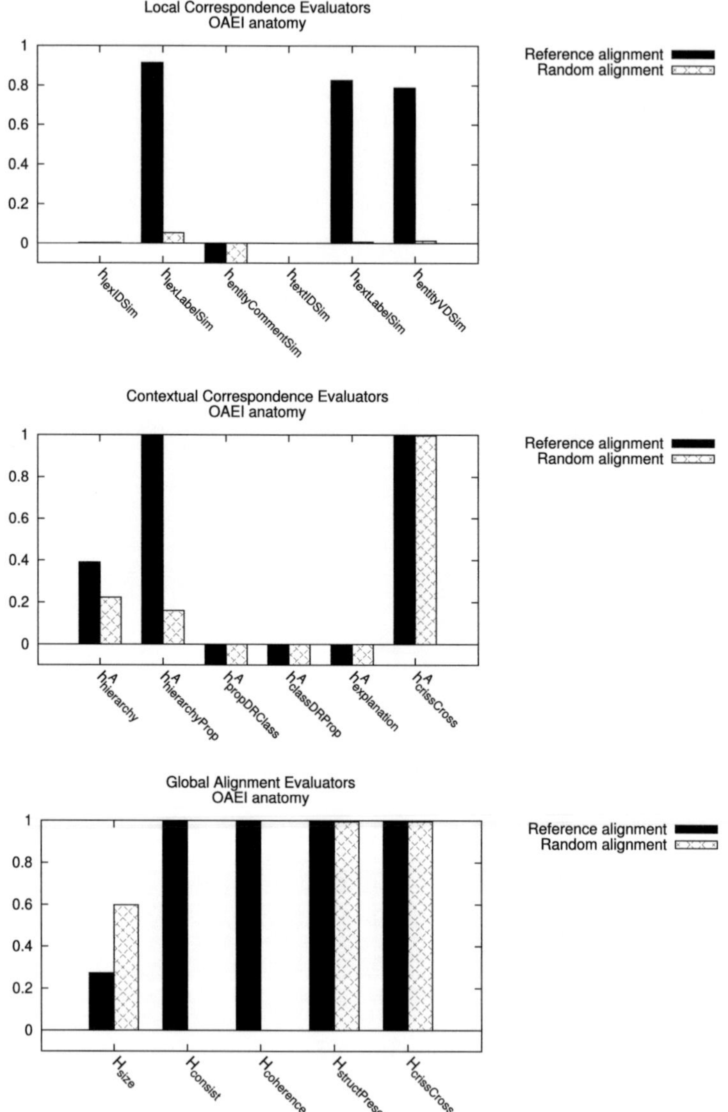

Figure 7.6.: Evaluation scores for different evaluator types for reference and random alignments of the two ontologies from the OAEI *anatomy* track. Inapplicability of an evaluator is indicated by a negative score.

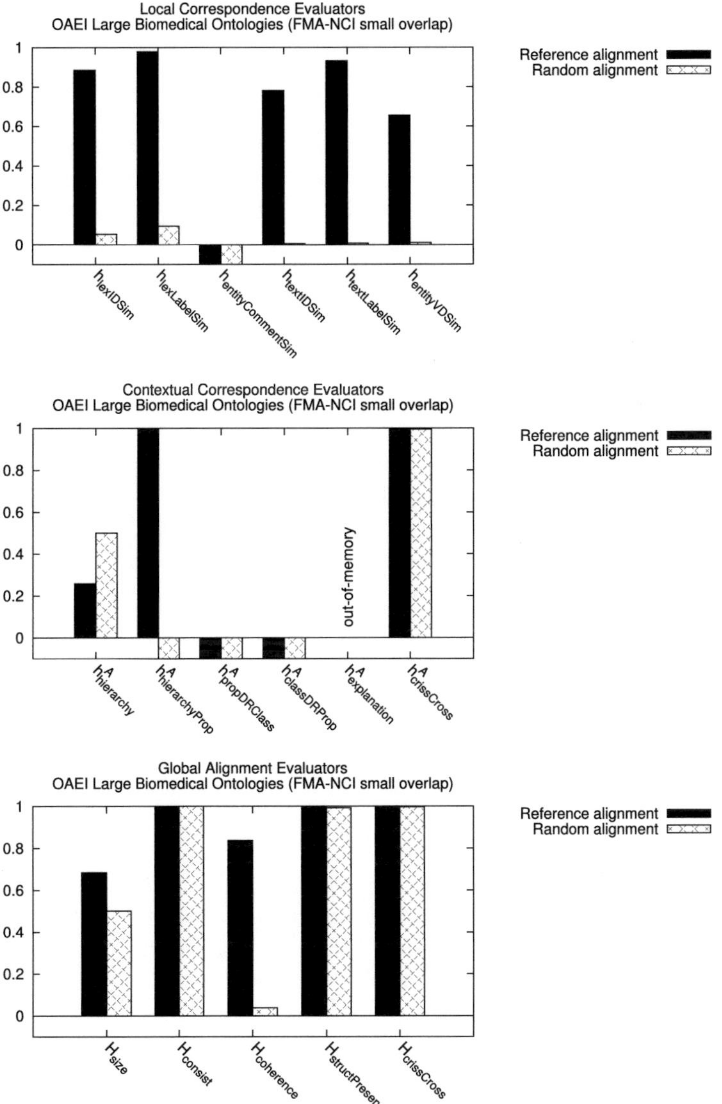

Figure 7.7.: Evaluation scores for different evaluator types for reference and random alignments of the two ontologies from the OAEI *large biomedical ontologies* track (small overlap). Inapplicability of an evaluator is indicated by a negative score.

7.3.3. Discussion

It can be observed that the structural evaluators $h^A_{\mathrm{crissCross}}$, $H_{\mathrm{structPreserv}}$, and $H_{\mathrm{crissCross}}$ do not show any significant discriminatory behaviour for all considered test cases. This is most likely due to the structural constitution of the ontologies in these test cases, which presumably reflect the constitution of most real-world ontologies in this respect. Having a closer look reveals that the metric according to Joslyn *et al.* [71] requires common "paths" on the subsumption hierarchy of entities in any two correspondences in order to cause the metric to produce a low score. Any correspondences with entities that have no common subsumer score high. In ontologies that are relatively shallow in terms of their subsumption hierarchy compared to their total number of entities, a large fraction of correspondences do not have entities with common subsumers. Thus the metric provides high evaluation scores even for random alignments. A similar phenomenon can be observed for the criss-cross evaluation metrics both on the correspondence and on the alignment level. Correspondences of entities that are not on the same path in terms of super- or subentities cannot cross and thus cannot produce a low evaluation score.

It should be noticed that the observations regarding the criss-cross metric on the correspondence level are an aggregated value. This does not mean that individual correspondences are not evaluated low if they are crossing other correspondences. So for all algorithm internal decisions that depend on single correspondence evaluations as a support heuristic, the criss-cross metric can provide valuable information. For the criss-cross metric on the alignment level, as well as the structural preservation metric, the evaluation scores do indeed provide no significant information regarding the quality of an alignment. In the case of the structural preservation metric, the computation of Joslyn *et al.* might require adjustments in order to account for the structural properties of ontologies. Additionally, it might be a promising approach to convert the structural preservation metric from a global alignment level metric to a contextual correspondence level metric in order to achieve a fine-grained discrimination for correspondences similar to the criss-cross evaluator.

A delicate metric is the alignment size evaluator H_{size}. The figures show that for all randomly generated alignments the size evaluator scores about 0.5. This is due to the way random alignments are generated here: each possible correspondence is selected for the alignment with a probability of 0.5. The motivation of the alignment size evaluator is to reward larger alignments in order to maximise the overlap. As it can be observed in Figures 7.5 and 7.6 for the *conference* and *anatomy* track, respectively, some ontologies have a small overlap. If the influence of the alignment size evaluator is too high, *e.g.* in a weighted average aggregation (cf. Section 4.2.2), resulting alignments tend to be too large, which causes the recall score to decrease (cf. Sections 7.2.3 and 7.2.4).

For all test ontology pairs used in this study the random baseline alignments always induce an inconsistency, apart from the *large biomedical ontologies* test case. Since inconsistency implies that all classes are unsatisfiable, the coherency evaluator $H_{\mathrm{coherence}}$ delivers 0. As a consequence the explanation-based evaluator $h^A_{\mathrm{explanation}}$ on the correspondence level becomes inapplicable for an inconsistent ontology. These results corroborate Conjecture 3 from Section 1.1 by demonstrating that even for seemingly inexpressive and simple ontologies such as the OAEI *benchmarks*, inconsistencies are induced by any of five randomly created alignments. It has to be noted that a merged ontology based on an alignment being inconsistent causes the explanation evaluator to fail quickly. In the single case of *large biomedical ontologies* where the merged ontology based on the alignment is not inconsistent, the explanation evaluator fails with an out-of-memory error. The reason for this behaviour is the great computational expense that comes with reasoning intensive calculations. This makes the explanation evaluator $h^A_{\mathrm{explanation}}$ difficult to employ even for moderately sized ontologies. The explanation computation could be optimised for the special case considered in ontology alignment[19].

7.4. Convergence and Anytime Alignment

This section analyses the convergence of the MapEVO and MapPSO prototypes. An experiment has been conducted using the OAEI *benchmarks* test case 103. The insights from the previous Section 7.3 are that the quality of the result produced by the introduced algorithms strongly depends on the suitability of the evaluation metrics used to compose the objective function. For the *benchmarks* test case 103 an objective function can be configured such that the algorithm provides good results in terms of precision, recall, and F-measure.

---

[19]The black box simple expand-shrink strategy [72] could be modified in order to specifically check the axioms induced by the correspondences in the alignment, thus avoiding the computation of complete explanations.

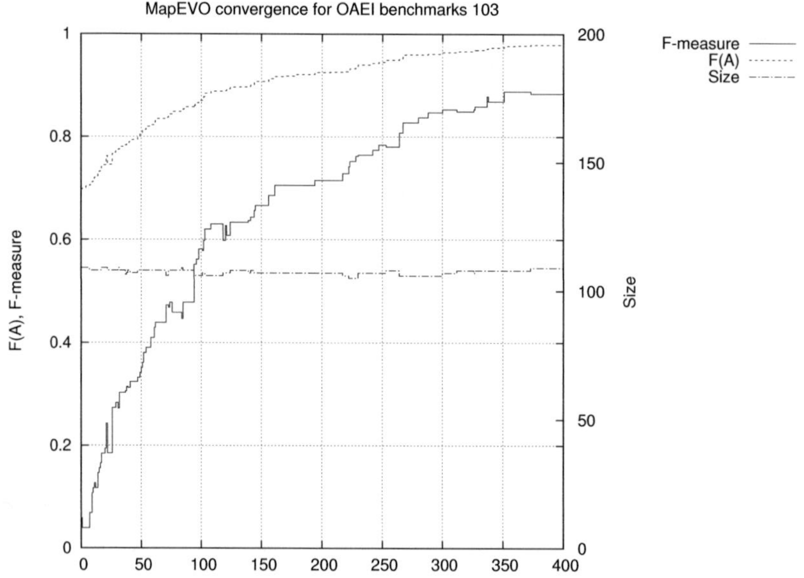

Figure 7.8.: Convergence of MapEVO for the *benchmarks* test case 103 regarding alignment quality (objective function) $F(A)$, alignment size, and F-measure (with respect to the provided reference alignment).

Figures 7.8 and 7.9 illustrate the convergence behaviour of MapEVO and MapPSO, respectively, in terms of the objective function (alignment quality function $F(A)$), alignment size, as well as the F-measure evaluation of the intermediate alignments with respect to the reference alignment. Quality, size, and F-measure refer to the best alignment with respect to $F(A)$ in the whole population in each iteration.

For both experiments, the same configuration[20] was used. In particular, both algorithms were configured with a population size of 40 and an objective function comprising the evaluation metrics H_{size}, $H_{\text{crissCross}}$, and $H_{\text{structPreserv}}$ on the alignment level, as well as h_{lexIDSim}, $h_{\text{lexLabelSim}}$, $h_{\text{textLabelSim}}$, $h_{\text{entityVDSim}}$, $h^{A}_{\text{hierarchy}}$, $h^{A}_{\text{hierarchyProp}}$, and $h^{A}_{\text{crissCross}}$ on the correspondence level.

---

[20]Since the two algorithms have different configuration parameters, using the same configuration in this experiment refers to the same configuration of *shared* components, *i.e.* population size and objective function (cf. Section 6.2).

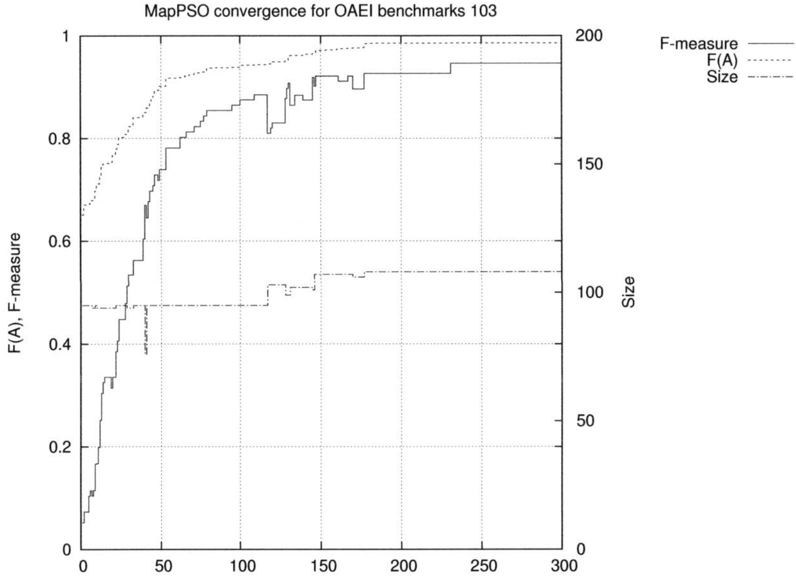

Figure 7.9.: Convergence of MapPSO for the *benchmarks* test case 103 regarding alignment quality (objective function) $F(A)$, alignment size, and F-measure (with respect to the provided reference alignment).

It can be observed that the alignment quality is continuously improving throughout the iterations. However, an improvement of the alignment quality used as objective function does not always directly correlate with an improvement of F-measure. This shows that the objective function does not always correctly reflect the quality criteria expected by the reference alignment. In most cases a drop in F-measure is correlated with a change in the alignment size. Increasing the size of an alignment is considered desirable by the objective function due to the impact of H_{size} that is striving for larger alignments. On the other hand, a larger alignment bears the risk of containing more suboptimal correspondences, which is penalised by the precision evaluation and thus influences the F-measure.

This behaviour is relevant for Conjecture 4 from Section 1.1 regarding the anytime behaviour of biologically-inspired optimisation techniques. In fact, interrupting the optimisation run at any time indeed delivers an alignment that is the best one found so far according to the objective function. However, it is not guaranteed that this alignment is the best one found so far compared with a reference alignment. Nevertheless, both Figures 7.8 and 7.9 show a relatively monotonic increase of F-measure such that the general claim can be defended that the presented algorithms demonstrate anytime behaviour in order to give up quality for runtime.

Comparing the convergence plots for MapEVO and MapPSO points out a slightly different convergence speed for both prototypes. Thereby, MapPSO demonstrates faster convergence, and reaches a stable state after about 250 iterations. In contrast, MapEVO converges slower and reaches a stable state after about 400 iterations. This difference is strongly influenced by algorithm specific parameters such as for instance the probability bounds or the swap frequency delimiter ϱ for MapEVO, or β, γ, λ, κ, and σ for MapPSO.

7.5. Case Study: Gene Ontology (GO) and Medical Subject Headings (MeSH)

The Gene Ontology (GO) [125] is a controlled and well-maintained vocabulary for annotating genes and gene products. It describes so-called "terms" in three main branches, namely *cellular component* (2,980 terms), *biological process* (22,382 terms), and *molecular function* (9,329 terms)[21] [53].

The Medical Subject Headings (MeSH) thesaurus contains 26,142 descriptors[22] about the domains of *medicine, nursing, dentistry, veterinary medicine, health care systems,* and *preclinical sciences* [53]. One of its use cases is to serve as an index for about five thousand biomedical journals in the MEDLINE / PubMED® database[23].

Using the GO and MeSH in an integrated manner can provide an added value to biomedical information systems [27]. Motivated by a requirement from the *THESEUS* SME 2009 award winning project GoOn[24] the

---

[21]The numbers refer to the ontology version 1.2830, dated 30/03/2012 16:15 `http://www.geneontology.org/GO.downloads.ontology.shtml` (accessed March 31, 2012)

[22]`http://www.nlm.nih.gov/pubs/factsheets/mesh.html` (accessed April 2, 2012)

[23]`http://www.ncbi.nlm.nih.gov/pubmed/` (accessed April 2, 2012)

[24]`http://theseus-programm.de/en/942.php` (accessed April 3, 2012)

MapPSO prototype was used for computing an alignment between GO and MeSH.

For the experiment, the OWL versions of the ontologies were used, where the GO contains ~31,650 classes, and the MeSH contains ~15,340 classes. The experiment [19] has been executed using the Amazon Web Services™ (AWS) cloud infrastructure using the cloud deployment mechanisms [17] presented in Section 6.5.3.

The cloud infrastructure was configured such that a single particle is evaluated on each Amazon Elastic Compute Cloud (EC2) instance. The chosen EC2 instance type was "m1.small" for all instances, which is a 32 bit single-core compute node with 1 EC2 Compute Unit, which "provides the equivalent CPU capacity of a 1.0-1.2 GHz 2007 Opteron or 2007 Xeon processor"[25] with 1.7 GB memory, operated by Linux (cf. Table 7.8).

Since no reference alignment is available for these data sets, no thorough evaluation of the alignment quality could be provided. However, the convergence of the algorithm could be measured in terms of the fitness scores of each particle throughout the iterations. Figure 7.10 shows the convergence behaviour regarding alignment quality of the MapPSO algorithm for this experiment, while Figure 7.11 shows the convergence behaviour regarding alignment size. The particles are depicted by the different lines in the figures.

MapPSO was configured to run with 8 particles and 30 iterations using an asynchronous communication strategy (cf. Section 6.5.3), which is reflected by the varying time intervals between iterations for the different particles. It can be observed that the wall-clock time taken for an iteration strongly depends on the size of the alignment represented by the particle, since larger particles have longer runtimes. For instance, the initially largest particle according to Figure 7.11 takes the longest time for 30 iterations. Moreover, it can be observed that the instant a particle adjusts its size typically coincides with a significant improvement of its quality.

No other state-of-the-art ontology alignment system is reported to be capable of exploiting distributed computing architectures or cloud infrastructures. For this reason it is difficult to carry out a fair comparative study on how well other state-of-the-art can cope with the same data set. While in the experiment, MapPSO was given 8 instances of relatively small EC2 nodes, the other systems were given a single, but more powerful machine. Table 7.8 contrasts the two settings.

---

[25]http://aws.amazon.com/ec2/instance-types/
(accessed April 11, 2012)

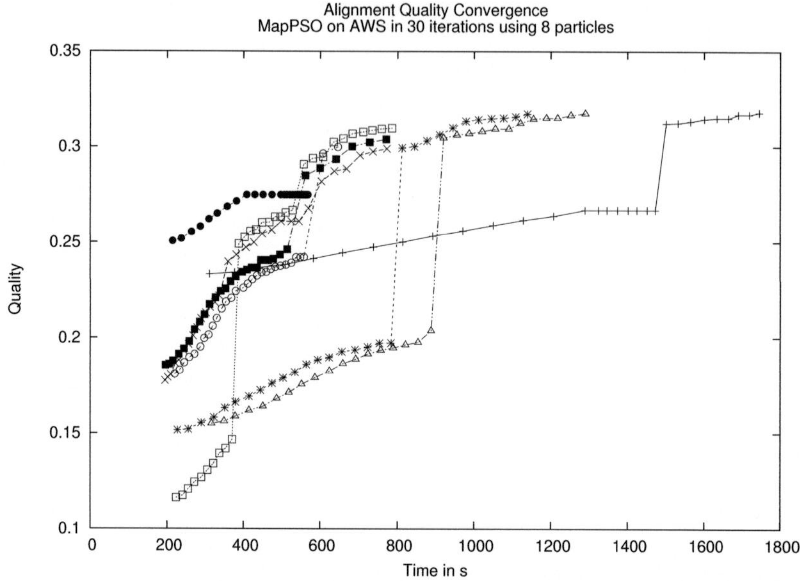

Figure 7.10.: Convergence behaviour of the MapPSO algorithm regarding alignment quality (objective function) on the AWS cloud infrastructure when aligning the Gene Ontology (GO) with the Medical Subject Headings (MeSH). Each line depicts the progress of a particle and line marks denote iterations.

In the experiment only those alignment systems were tested that could successfully process the OAEI 2011 *anatomy* track. Table 7.9 shows the results of the experiments. It can be observed that the MaasMatch system [114] is the only system that could provide a non-empty alignment in about 17 hours. This was only possible when providing 4 GB of heap space, since the system also failed with 2 GB as used on a single compute node for MapPSO. The success of MaasMatch may be related to the fact that this was the only system running multi-threaded and thus making full use of the 4 cores available on the test machine.

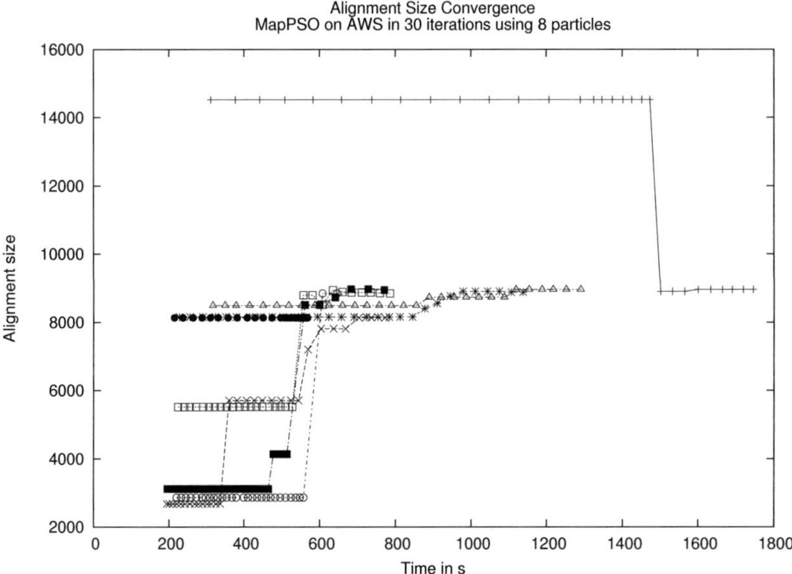

Figure 7.11.: Convergence behaviour of the MapPSO algorithm regarding alignment size on the AWS cloud infrastructure when aligning the Gene Ontology (GO) with the Medical Subject Headings (MeSH). Each line depicts the progress of a particle and line marks denote iterations.

7.6. Discussion

The suitability of biologically-inspired optimisation techniques for solving the ontology alignment problem has been shown by several experiments. It could be demonstrated that biologically-inspired optimisation techniques provide an answer to all conjectures declared in Section 1.1.

The quality of the results obtained by the two presented algorithms was measured using the Ontology Alignment Evaluation Initiative (OAEI) data sets. Since 2008, the MapPSO system constantly participated in the official campaign. It could be observed that the results are of diverse quality. The MapPSO system performed competitively in the *directory* track. While for some test cases in the *benchmarks* track very good results with respect to classical precision and recall measures could be obtained, there were other cases, where the result quality was poor. Interestingly, for

Table 7.8.: Computing infrastructure used for experiments with MapPSO and other systems for aligning the Gene Ontology (GO) with Medical Subject Headings (MeSH). MapPSO used 8 EC2 instances on the Amazon® Web Services (AWS) cloud.

	AWS EC2 instance (MapPSO)	Local machine (other systems)
CPU	1×1 ECU (~ 2007 Opteron/Xeon 1.0-1.2 GHz)	$4 \times$ Intel® Core™ i7 CPU, M620, 2.67 GHz
Memory	1.7 GB	4 GB
Architecture	i386	x86
Operating System	32 bit Linux	64 bit Linux

both MapPSO and MapEVO it was reported by the organisers that with respect to the generalised precision and recall measures (cf. Section 7.1) a significant improvement can be observed. This matches the expectation that biologically-inspired optimisation techniques provide a coarse-grained search, and usually require local search techniques to fine-tune the results [106, Sect. 6.3].

A second aspect regarding the result quality is its correlation with the objective function. Those test cases, where good results could be obtained are those for which the similarity metrics used for instantiating the objective function show a high discriminative behaviour (cf. Section 7.3). This means that similarity metrics are used that can clearly distinguish good alignments from bad ones, high quality results can be obtained. This supports the main argument that biologically-inspired optimisation techniques can be used for solving the ontology alignment problem. However, a high quality objective function is required that encodes mechanisms to exploit the relevant ontology characteristics. In order to improve the results regarding alignment quality, developing additional metrics that cover further ontology characteristics would be beneficial, as well as improving the existing metrics. Examples for possible improvements would be the adaptation of the explanation evaluator or the structural preservation evaluator as discussed in Section 7.3.3, or the interpretation of complex entity expressions in most of the contextual evaluators.

Independent of how the objective function exploits the relevant ontology characteristics, convergence could be observed for both MapEVO and

Table 7.9.: OAEI 2011 participants used for aligning the Gene Ontology (GO) with Medical Subject Headings (MeSH). Only those systems were tested that could successfully process the OAEI 2011 *anatomy* track.

System	Description
AgrMaker	*system was not made available for testing*
Aroma	empty alignment after ∼20 seconds
CSA	empty alignment after ∼7 minutes (`ArrayIndexOutOfBoundsException`)
CODI	internal error (`IllegalStateException`)
Lily	out-of-memory
LogMap	out-of-memory
MaasMatch	*non-empty alignment after ∼17 hours*

MapPSO. In particular this convergence could be observed for aligning large biomedical ontologies (Gene Ontology and Medical Subject Headings) on a cloud infrastructure (cf. Sections 7.4 and 7.5). The studies show the expected anytime behaviour, where at any point in time during the optimisation run, the currently best alignment can be obtained. A correlation between F-measure of this intermediate alignment and the objective function value could be observed. The experiments further demonstrate that large ontologies can be processed at all, and distributed computing resources, *e.g.* cloud infrastructures, can be used. Insights obtained from the OAEI 2011.5 *large biomedical ontologies* track, however, demonstrate that some state-of-the-art alignment tools are able to process large ontologies, as well. Many of those systems, however, require powerful monolithic computing resources, whereas MapPSO can utilise a large number of individually less powerful resources. This goes along with the paradigm shift from monolithic to distributed computing infrastructures, realised, for instance, by modern cloud computing solutions.

Evaluation in the *SEALS* Platform. The *SEALS* project has the goal to provide a universal evaluation platform that allows for controlled and repeatable execution of tools. The evaluation environment is set up as a virtual machine in which the tool under evaluation is automatically deployed and executed. While this environment is suitable and desired for executing and comparing monolithic systems that are designed to run on a single machine, it is unsuitable for algorithms that are designed to be executed in a distributed environment, such as cloud infrastructures. Due to this limitation, the MapPSO and MapEVO algorithms cannot use their full potential when being executed in the *SEALS* platform. The OAEI tracks providing large ontologies to be aligned thus could only be addressed using a configuration that uses the computing resources available on a single compute node. For instance, in the case of MapPSO this means that only a few particles can be used and since all particles need to be computed on a single compute node, the runtime increases drastically for one iteration, and consequently for the overall computation.

Another criticism regarding the evaluation in the *SEALS* platform is the enforced single-configuration policy, meaning that participating alignment systems have to use a single configuration for all OAEI tracks or have to implement self-tuning. This is done by some systems, in the case of AgreementMaker even to an extent that is discouraged by the OAEI organisers. They report that "AgreementMaker used machine learning techniques to choose automatically between one of three settings optimized for the benchmark, anatomy and conference data set. It used a subset of the available reference alignments as input to the training phase and clearly a specific tailored setting for passing these tests. This is typically prohibited by OAEI rules. However, at the same time, Agreement-Maker has improved its results over last year so we found interesting to report them" [47].

Given the different characteristics of ontologies in the different OAEI tracks, what the OAEI campaign is essentially evaluating is the ability of alignment systems to automatically adapt their configuration. Neither MapPSO nor MapEVO implement sophisticated self-adaptation procedures. However, a very simple preprocessing was implemented for the second *SEALS* evaluation campaign in order to have a sufficient number of iterations in order to achieve convergence for each test case (*large biomedical ontologies* excluded due to time restrictions).

8. Conclusion

An investigation of how biologically-inspired optimisation methods can be used for solving the ontology alignment problem was presented in this book. In particular two approaches were introduced, an Evolutionary Algorithm, and a Discrete Particle Swarm Optimisation algorithm, both adapted for and applied to the ontology alignment problem. Biologically-inspired optimisation techniques demonstrate several interesting features and properties that suite well some of the challenges faced in the problem domain of ontology alignment. More precisely the aspects considered here were laid out in Section 1.1 in terms of four conjectures:

- Scalability improvement of ontology alignment due to the inherent parallelisability of population-based biologically-inspired optimisation techniques.

- Adaptability of the alignment algorithm to ontologies with different modelling characteristics due to the independence of biologically-inspired optimisation techniques from their objective function.

- Consideration of global alignment evaluation criteria, as well as other inter-correspondence dependencies due to the global representation and assessment of solutions in biologically-inspired optimisation techniques.

- Support for a gradual trade-off between alignment quality and run-time due to the inherent anytime behaviour of biologically-inspired optimisation techniques based on their iterative execution.

This chapter summarises the main findings in Section 8.1 and provides an outlook on future research directions in Section 8.2.

8.1. Results

Two novel algorithms were developed in this book to solve the ontology alignment problem:

- An Evolutionary Algorithm based on ideas from Evolutionary Programming and Extremal Optimisation, as well as other facets that have been shown useful in related applications. The algorithm manages a population of species, each undergoing adapted mutations in every iteration. In frequent intervals, a ranked-based selection is done removing the worst species and allowing the best ones to reproduce. The mutation operators modifying each individual's configuration are influenced by correspondence confidences as support heuristic. Furthermore they are influenced by the iteration progress to limit the amplitude of changes towards the end of the optimisation run.

- A Discrete Particle Swarm Optimisation Algorithm inspired by an approach to solving the structurally similar problem of attribute selection for a machine learning classifier. The algorithm maintains a population of particles, which are moving through the problem space of all possible candidate alignments. The new position of each particle, *i.e.* the set of correspondences it represents, is influenced by two factors:

    - Each correspondence is likely to be preserved in the alignment if it is also present in the particle's personal best configuration, or in the best configuration of any particle in the particle's neighbourhood.

    - For each correspondence its confidence, as a support heuristic, influences the decision whether it will be preserved in the next iteration.

In order to compose useful objective functions, a collection of similarity metrics has been provided. Three types of metrics were presented:

- Local correspondence level metrics, solely exploiting information available for the two corresponding entities themselves and their ontology context.

- Contextual correspondence level metrics, exploiting the alignment context, *i.e.* the evaluation of correspondences in the presence of other correspondences.

- Global alignment level metrics, considering criteria accessible only for complete alignments.

The presented similarity metrics can be seen as an exemplary tool box in order to instantiate arbitrary objective functions. For special alignment scenarios, additional metrics are required, which is why the presented evaluation metrics are not meant to be an exhaustive collection.

Regarding the motivational conjectures from Section 1.1, the developed algorithms show satisfactory behaviour, summarised in the following.

Scalability (Conjecture 1). There is a tendency of ontologies to become larger due to the increasing amount of knowledge and the availability of powerful information systems to exploit it. This can in particular be observed in the biomedical domain. Population-based biologically-inspired optimisation techniques are inherently parallelisable in the way that in each iteration, each population member evaluates a candidate solution independently. This property was preserved when adjusting the classical optimisation algorithms to the alignment problem in this book. The remaining bottleneck in the Evolutionary Algorithm is the centralised ranking and controlled duplication and removal of species from the population. In the Particle Swarm Optimisation algorithm, the bottleneck is the communication via social networks and, compared to the Evolutionary Algorithm, the more fine-grained exchange of information on the correspondence level (shared correspondences in particles).

On the implementation side these requirements regarding the communication bottleneck, particularly in distributed environments, were considered. On the one hand, a novel API, named *KADMOS*, was developed for the use in both prototypes with a particular focus on serialisability and a communication API for cloud infrastructures. On the other hand, for the MapPSO prototype, an asynchronous particle update strategy was applied, and a concept called *particle pooling* was introduced in order to improve parallel efficiency.

The applicability to large ontologies was demonstrated by computing an alignment for the Gene Ontology (GO) and the Medical Subject Headings (MeSH) with several tens of thousands of concepts. With the MapPSO prototype an alignment was computed on the Amazon Web Services^{TM} cloud infrastructure. Due to the absence of a suitable reference alignment, no quality evaluation of the results could be done, however, the convergence of the algorithm could be observed, which shows the continuous improvement of the computed alignment with respect to the objective function used.

A recent alignment system evaluation and comparison in the context of the Ontology Alignment Evaluation Initiative revealed that some other state-of-the-art systems are also able to process large ontologies. In particular the availability of a powerful (high-memory) computing resource enables some systems to cope with large inputs. This leads to a differentiated view of the advantage in scalability of biologically-inspired optimisation techniques applied to ontology alignment. However, in the light of the paradigm shift towards distributed and cloud-based computing infrastructures the parallelisability of the presented algorithms can be seen as an advantage nonetheless. For the population-based approaches the single compute nodes in the distributed infrastructure can be relatively moderate in terms of number of CPUs or cores, and memory, which is compensated by the total number of compute nodes utilised.

Flexibility (Conjecture 2). Different interpretations of the term "ontology", the availability of ontology languages of different expressiveness, as well as the different use cases for ontologies result in a variety of ontologies demonstrating different modelling characteristics. These characteristics play an important role when an alignment between two ontologies is to be computed. The notion of "optimality" for an ontology alignment depends on a ranking of candidate alignments, in order to make a statement such as "alignment A is better than alignment B". In biologically-inspired optimisation techniques, ranking criteria are encoded in an objective function, where the optimiser strives for finding the solution with the best objective function value.

The presented applications of biologically-inspired optimisation techniques for ontology alignment use an objective function for global alignment evaluations, and a support heuristic for evaluations on the correspondence level. Both are treated as black boxes by both optimisation metaheuristics. In this sense the presented algorithms are applicable for aligning ontologies of different characteristics, provided the way to exploit these characteristics is appropriately encoded in the objective function and support heuristic.

Regarding the implemented prototypes, both MapEVO and MapPSO utilise the same objective function and support heuristic. Various similarity metrics are provided as an external library, named *HARMONIA* Commons, making it possible to configure arbitrary objective functions and support heuristics, and facilitate extensibility.

Empirical studies have shown that the configuration of the objective function (and support heuristic) is crucial for obtaining high quality results. To this end, a correlation between the discrepancy of used similarity metrics and the result quality in various test cases was demonstrated. This correlation explains the diverse results of the presented implementation prototypes in the Ontology Alignment Evaluation Initiative (OAEI), where the same objective function was used for aligning ontologies with a wide range of modelling characteristics. In those cases, where the objective function and support heuristic were suitably configured, results of good quality could be obtained.

Global Alignment Metrics (Conjecture 3). In particular for those ontologies that exploit ontology language features of higher expressiveness, care must be taken when computing an alignment, in which correspondences are meant to be interpreted with an equivalence semantics. The reason is that treating correspondences as equivalence axioms can have an impact on the consistency or coherency of the merged ontology based on an alignment. Inter-correspondence effects can cause classes to become unsatisfiable, or even the merged ontology to become inconsistent. The way biologically-inspired optimisation techniques treat the objective function as a black box allows for encoding both inter-correspondence evaluation criteria (evaluation of a correspondence in the context of other correspondences), or arbitrary other global alignment metrics into the objective function and the support heuristic.

The *HARMONIA* Commons library is provided as a separate implementation module for being used by the two prototypes MapEVO and MapPSO, as well as by arbitrary other software modules. It allows for easy encoding of contextual correspondence level and global alignment level evaluation metrics.

Experiments with various similarity metrics have shown that even for seemingly simple ontologies, random alignments with their correspondences interpreted as equivalence axioms induce inconsistencies. This is a strong indication that dedicated similarity metrics are required in order to avoid inconsistency or incoherency inducing alignments.

Approximate and Anytime Alignment (Conjecture 4). Depending on the application scenario of ontology alignment, the expectation regarding result quality can vary. While safety-critical applications require high result quality, simple search applications might require less quality. Conversely, those simple search applications might require faster response time than critical applications, where long runtimes can be accepted. The iterative nature of biologically-inspired optimisation techniques bears the potential to interrupt the process at any point in time and retrieve the best intermediate result computed so far. This property was preserved when adjusting the classical optimisation algorithms to the alignment problem in this book, since implementing this feature is straightforward.

The convergence of the prototypes MapEVO and MapPSO in terms of the improved objective function value was compared with the F-measure denoting the quality of intermediate alignments with respect to a given reference alignment. In this empirical study, a correlation between objective function value and F-Measure could be observed[1].

One other interesting observation in the course of evaluating the approaches was that the results found by both MapPSO and MapEVO in the context of the Ontology Alignment Evaluation Initiative were often not exactly matching the reference alignment, but were close to the reference. This could have been shown by applying relaxed precision and recall metrics. It is a typical behaviour of biologically-inspired optimisation metaheuristics, that they provide coarse-grained near-optimal results, and typically require a local search component in order to find the absolute optimum [106].

8.2. Outlook

This work demonstrates a first systematic study of applying biologically-inspired optimisation techniques to directly solve the ontology alignment problem. Due to the numerous facets that can be focused on and varied when applying an optimisation method to a given problem, there are plenty of directions that can be followed in the future in order to extend and refine the work.

One research direction is clearly the improvement of alignment results. As the presented studies have shown, the result quality strongly depends on the instantiation of the objective function and thus on the feasibility

---

[1]The correlation between objective function value and F-measure was not entirely smooth. More precisely, there are cases where an improvement of the objective function score correlated with a decrease in F-measure, which was mostly due to a change in the size of the alignment in the respective iterations. As from previous experiments, the reason is in the definition and configuration of the objective function, in particular with respect to the influence of the alignment size.

of used similarity metrics. Developing new generic and special-purpose similarity metrics thus is always a way to improve result quality, for instance, in biomedical use cases where the UMLS metathesaurus [22] would be a valuable resource. However, in order to provide a generic zero-configuration alignment system, sophisticated preprocessing techniques have to be applied in order to self-adapt the objective function and other critical parameters. Successful alignment systems with respect to the OAEI data sets perform such a self-adaptation. AgreementMaker, for instance, uses a mechanism of base matcher self-assessment in order to estimate the "local confidence" [34] of selected correspondences. The RiMOM system features the dynamic weight calculation for the predicting value aggregation based on the analysis of ontology characteristics in a preprocessing step [85, 134]. Another option for improving the objective function on demand would be to incorporate interactive components, such as user interaction for adjusting parameters [127], or directly influencing the optimisation algorithm [63].

In case the algorithm will be applied in a fixed, well-known use case domain, *a priori* configurations of the objective function would be beneficial. Using the methods of Martinez-Gil [86, 87] (GOAL), or Ritze and Paulheim [110] (ECOMatch), one could obtain such a configuration in case a partial reference alignment is available or manually created upfront. Similarly, test planning methods, or Genetic Programming [79] could be applied for finding a good fitness function.

Another direction of future research that is likely to have an impact on improving the quality of generated alignments is the application of local search components, since optimisation metaheuristics are typically only providing only near optimal solutions [106]. Such local search components could be applied both during the optimisation run, or as a postprocessing step at the end.

Obviously, an exhaustive in-depth study of all available biologically-inspired optimisation metaheuristics with respect to their applicability to the ontology alignment problem was beyond the scope of this book. However, for instance, using Ant Colony Optimisation could be another promising approach for solving the ontology alignment problem. Similar to the metaheuristics used in this book, the individuals (ants) could be computed independently from each other. However, a shared data structure would be necessary for their communication via stigmergy. This requires a different and more complex deployment in distributed computing infrastructures, and could hamper scalability without efficient optimisations on the implementation side. The presented Evolutionary Algorithm does not exploit the feature of Self-Organised Criticality according to the Bak-Sneppen model [5], since it requires the notion of a global neighbourhood relation of individuals in the population. This in turn requires frequent pairwise comparisons of individuals on their component level (in case shared solution components are used to determine closest neighbours, as proposed by Randall [105]). This consequently imposes additional challenges for deployment on distributed computing infrastructures, as it is the case for Ant Colony Optimisation.

Further extension of the presented approaches would be to relax the restrictions of ontology alignment representations, in order to allow for entities corresponding to more than one other entity. This could be required in some use cases, however, such a relaxation would significantly increase the solution space to $2^{\#O_1 \cdot \#O_2}$ making the optimisation problem even more difficult. Another related approach that might be worth exploring is to allow the algorithm to move candidate solutions through infeasible solution spaces during the optimisation run. This could make it easier to detect better correspondences for entities which are already participating in another correspondence. Currently, in an iteration no new correspondence can be created for an entity, which is "blocked" by participating in another correspondence, since otherwise the alignment would become invalid.

In cases where ontology alignments have to be computed in changing environments, *i.e.* with frequently or constantly changing ontologies, biologically-inspired optimisation techniques can provide a valuable solution [62]. The way the presented approaches are useful in such scenarios, could be analysed in future studies.

Last but not least, the possibility to exploit Infrastructure-as-a-Service (IaaS) cloud services to build a scalable and elastic Software-as-a-Service (SaaS) alignment service opens the door for business models providing flexible alignment services for the demands of future semantic applications.

Bibliography

[1] Gene M. Amdahl. Validity of the Single Processor Approach to Achieving Large Scale Computing Capabilities. In *Proceedings of the AFIPS '67 (Spring) Joint Computer Conference*, volume 30 of *AFIPS Conference Proceedings*, pages 483–485, Washington, DC, USA, April 1967. Thompson Books.

[2] F. Baader, D. Calvanese, D. L. McGuinness, D. Nardi, and P. F. Patel-Schneider, editors. *The Description Logic Handbook: Theory, Implementation, and Applications*. Cambridge University Press, New York, NY, USA, 2003.

[3] Thomas Bäck, David B. Fogel, and Zbigniew Michalewicz, editors. *Evolutionary Computation 1: Basic Algorithms and Operators*. Taylor & Francis, New York, NY, USA, 2000.

[4] Thomas Bäck, David B. Fogel, and Zbigniew Michalewicz, editors. *Evolutionary Computation 2: Advanced Algorithms and Operators*. IOP Publishing, Philadelphia, PA, USA, 2000.

[5] Per Bak and Kim Sneppen. Punctuated equilibrium and criticality in a simple model of evolution. *Physical Review Letters*, 71(24):4083–4086, December 1993.

[6] Sean Bechhofer and Alistair Miles. SKOS Simple Knowledge Organization System Reference. W3C recommendation, W3C, August 2009. http://www.w3.org/TR/2009/REC-skos-reference-20090818/.

[7] Zohra Bellahsene, Angela Bonifati, Fabien Duchateau, and Yannis Velegrakis. On Evaluating Schema Matching and Mapping. In Zohra Bellahsene, Angela Bonifati, and Erhard Rahm, editors, *Schema Matching and Mapping*, Data-Centric Systems and Applications, chapter 9, pages 253–291. Springer, Berlin, Heidelberg, 2011.

[8] Zohra Bellahsene, Angela Bonifati, and Erhard Rahm, editors. *Schema Matching and Mapping*. Data-Centric Systems and Applications. Springer, Berlin, Heidelberg, 2011.

[9] Christian Bizer, Tom Heath, and Tim Berners-Lee. Linked Data – The Story So Far. *International Journal on Semantic Web and Information Systems*, 5(3):1–22, 2009.

[10] Christian Blum and Andrea Roli. Metaheuristics in Combinatorial Optimization: Overview and Conceptual Comparison. *ACM Computing Surveys*, 35(3):268–308, September 2003.

[11] Jürgen Bock. Ontology Merging using Answer Set Programming and WordNet. Bachelor honours thesis, Faculty of Engineering and Information Technology, School of Information and Communication Technology, Griffith University, Nathan Campus, Brisbane, Australia, October 2006.

[12] Jürgen Bock. MapPSO Results for OAEI 2010. In Pavel Shvaiko, Jérôme Euzenat, Fausto Giunchiglia, Heiner Stuckenschmidt, Ming Mao, and Isabel Cruz, editors, *Proceedings of the 5th International Workshop on Ontology Matching (OM-2010)*, volume 689, pages 179–185, http://ceur-ws.org, November 2010. CEUR Workshop Proceedings.

[13] Jürgen Bock, Carsten Dänschel, and Matthias Stumpp. MapPSO and MapEVO results for OAEI 2011. In Pavel Shvaiko, Jérôme Euzenat, Tom Heath, Christoph Quix, Ming Mao, and Isabel Cruz, editors, *Proceedings of the 6th International Workshop on Ontology Matching*, volume 814, pages 179–183, http://ceur-ws.org, October 2011. CEUR Workshop Proceedings.

[14] Jürgen Bock, Peter Haase, Qiu Ji, and Raphael Volz. Benchmarking OWL Reasoners. In Frank van Harmelen, Andreas Herzig, Pascal Hitzler, Zuoquan Lin, Ruzica Piskac, and Guilin Qi, editors, *Proceedings of the ARea2008 Workshop*, volume 350, http://ceur-ws.org, June 2008. CEUR Workshop Proceedings.

[15] Jürgen Bock and Jan Hettenhausen. MapPSO Results for OAEI 2008. In Pavel Shvaiko, Jérôme Euzenat, Fausto Giunchiglia, and Heine Stuckenschmidt, editors, *Proceedings of the 3rd International*

Workshop on Ontology Matching (OM-2008), volume 431, pages 176–181, `http://ceur-ws.org`, October 2008. CEUR Workshop Proceedings.

[16] Jürgen Bock and Jan Hettenhausen. Discrete Particle Swarm Optimisation for Ontology Alignment. *Information Sciences*, 192:152–173, 2012.

[17] Jürgen Bock, Alexander Lenk, and Carsten Dänschel. Ontology Alignment in the Cloud. In Pavel Shvaiko, Jérôme Euzenat, Fausto Giunchiglia, Heiner Stuckenschmidt, Ming Mao, and Isabel Cruz, editors, *Proceedings of the 5th International Workshop on Ontology Matching (OM-2010)*, volume 689, pages 73–84, `http://ceur-ws.org`, November 2010. CEUR Workshop Proceedings.

[18] Jürgen Bock, Peng Liu, and Jan Hettenhausen. MapPSO Results for OAEI 2009. In Pavel Shvaiko, Jérôme Euzenat, Fausto Giunchiglia, Heiner Stuckenschmidt, Natasha Noy, and Arnon Rosenthal, editors, *Proceedings of the 4th International Workshop on Ontology Matching (OM-2009)*, volume 551, pages 193–199, `http://ceur-ws.org`, October 2009. CEUR Workshop Proceedings.

[19] Jürgen Bock, Sebastian Rudolph, and Michael Mutter. More than the Sum of its Parts – Holistic Ontology Alignment by Population-Based Optimisation. In Thomas Lukasiewicz and Attila Sali, editors, *Proceedings of the Seventh International Symposium on Foundations of Information and Knowledge Systems (FoIKS)*, volume 7153 of *Lecture Notes in Computer Science*, pages 72–91, Berlin, Heidelberg, March 2012. Springer.

[20] Jürgen Bock, Rodney Topor, and Raphael Volz. Ontology Merging using Answer Set Programming and Linguistic Knowledge. In Pavel Shvaiko, Jérôme Euzenat, Fausto Giunchiglia, and Bin He, editors, *Proceedings of the 2nd International Workshop on Ontology Matching*, volume 304, pages 301–305, `http://ceur-ws.org`, November 2007. CEUR Workshop Proceedings.

[21] Oliver Bodenreider, Terry F. Hayamizu, Martin Ringwald, Sherri De Coronado, and Songmao Zhang. Of Mice and Men: Aligning Mouse and Human Anatomies. In *Proceedings of the AMIA 2005 Annual Symposium*, pages 61–65, Bethesda, MD, United States, October 2005. American Medical Informatics Association.

[22] Olivier Bodenreider. The Unified Medical Language System (UMLS): integrating biomedical terminology. *Nucleic Acids Research*, 32(1):267–270, 2004.

[23] Stefan Boettcher and Allon G. Percus. Extremal Optimization: Methods derived from Co-Evolution. In Wolfgang Banzhaf, Jason M. Daida, A. E. Eiben, Max H. Garzon, Vasant Honavar, Mark J. Jakiela, and Robert E. Smith, editors, *Proceedings of the Genetic and Evolutionary Computation Conference*, volume 1, pages 825–832, San Francisco, CA, USA, July 1999. Morgan Kaufmann.

[24] Stefan Boettcher and Allon G. Percus. Nature's way of optimizing. *Artificial Intelligence*, 119(1–2):275–286, May 2000.

[25] Paolo Bouquet, Luciano Serafini, and Stefano Zanobini. Semantic coordination: a new approach and an application. In Dieter Fensel, Katia P. Sycara, and John Mylopoulos, editors, *Proceedings of the 2nd International Semantic Web Conference*, volume 2870 of *Lecture Notes in Computer Science*, pages 130–145, Berlin, Heidelberg, October 2003. Springer.

[26] Don Box, David Ehnebuske, Gopal Kakivaya, Andrew Layman, Noah Mendelsohn, Henrik Frystyk Nielsen, Satish Thatte, and Dave Winer. Simple Object Access Protocol (SOAP) 1.1. W3C recommendation, W3C, May 2000. http://www.w3.org/TR/SOAP.

[27] Anders Bresell, Bo Servenius, and Bengt Persson. Ontology annotation treebrowser: an interactive tool where the complementarity of medical subject headings and gene ontology improves the interpretation of gene lists. *Applied Bioinformatics*, 5(4):225–236, 2006.

[28] Caterina Caracciolo, Jérôme Euzenat, Laura Hollink, Ryutaro Ichise, Antoine Isaac, Véronique Malaisé, Christian Meilicke, Juan Pane, Pavel Shvaiko, Heiner Stuckenschmidt, Ondřej Šváb-Zamazal, and Vojtěch Svátek. Results of the Ontology Alignment Evaluation Initiative 2008. In Pavel Shvaiko, Jérôme Euzenat, Fausto Giunchiglia, and Heiner Stuckenschmidt, editors, *Proceedings of the 3rd International Workshop on Ontology Matching (OM-2008)*, volume 431, pages 73–119, http://ceur-ws.org, October 2008. CEUR Workshop Proceedings.

[29] Kumar Chellapilla and Gary B. Fogel. Multiple sequence alignment using evolutionary programming. In *Proceedings of the 1999 Congress on Evolutionary Computation*, volume 1, pages 445–452, Washington, DC, USA, July 1999. IEEE Computer Society.

[30] Namyoun Choi, Il-Yeol Song, and Hyoil Han. A Survey on Ontology Mapping. *SIGMOD Record*, 35(3):34–41, September 2006.

[31] Oscar Corcho and Asunción Gómez-Pérez. A Roadmap to Ontology Specification Languages. In Rose Dieng and Olivier Corby, editors, *Proceedings of the 12th International Conference on Knowledge Engineering and Knowledge Management Methods, Models, and Tools*, volume 1937 of *Lecture Notes in Computer Science*, pages 80–96, Berlin, Heidelberg, October 2000. Springer.

[32] Elon S. Correa, Alex A. Freitas, and Colin G. Johnson. A New Discrete Particle Swarm Algorithm Applied to Attribute Selection in a Bioinformatics Data Set. In *Proceedings of the 8th Genetic and Evolutionary Computation Conference*, pages 35–42, New York, NY, USA, 2006. ACM.

[33] Elon S. Correa, Alex A. Freitas, and Colin G. Johnson. Particle Swarm and Bayesian Networks Applied to Attribute Selection for Protein Functional Classification. In *Proceedings of the 9th Genetic and Evolutionary Computation Conference*, pages 2651–2658, New York, NY, USA, 2007. ACM.

[34] Isabel F. Cruz, Flavio Palandri Antonelli, and Cosmin Stroe. Efficient Selection of Mappings and Automatic Quality-driven Combination of Matching Methods. In Pavel Shvaiko, Jérôme Euzenat, Fausto Giunchiglia, Heiner Stuckenschmidt, Natasha Noy, and Arnon Rosenthal, editors, *Proceedings of the 4th International Workshop on Ontology Matching*, volume 551, pages 49–60, `http://ceur-ws.org`, October 2009. CEUR Workshop Proceedings.

[35] Charles Darwin. *On the Origin of Species by Means of Natural Selection, or the Preservation of Favoured Races in the Struggle for Life*. John Murray, London, United Kingdom, November 1859.

[36] Dipankar Dasgupta and Zbigniew Michalewicz, editors. *Evolutionary Algorithms in Engineering Applications*. Springer, Berlin, Heidelberg, 1997.

[37] Jérôme David, Jérôme Euzenat, François Scharffe, and Cássia Trojahn dos Santos. The Alignment API 4.0. *Semantic Web – Interoperability, Usability, Applicability*, 2(1):3–10, 2011.

[38] Kathrin Dentler, Christophe Guéret, and Stefan Schlobach. Semantic Web Reasoning by Swarm Intelligence. In Achille Fokoue, Yuanbo Guo, and Thorsten Liebig, editors, *Proceedings of the 5th International Workshop on Scalable Semantic Web Knowledge Base Systems*, volume 517, pages 1–16, http://ceur-ws.org, October 2009. CEUR Workshop Proceedings.

[39] Marco Dorigo. *Optimization, Learning and Natural Algorithms*. PhD thesis, Dipartimento di Elettronica, Politecnico di Milano, Milan, Italy, 1992.

[40] Marco Dorigo and Luca Maria Gambardella. Ant Colony System: A Cooperative Learning Approach to the Traveling Salesman Problem. *IEEE Transactions on Evolutionary Computation*, 1(1):53–66, April 1997.

[41] Marco Dorigo and Thomas Stützle. Ant colony optimization: Overview and recent advances. In Michel Gendreau and Jean-Yves Potvin, editors, *Handbook of Metaheuristics*, volume 146 of *International Series in Operations Research & Management Science*, pages 227–263. Springer, Berlin, Heidelberg, 2nd edition, 2010.

[42] Marc Ehrig and Jérôme Euzenat. Relaxed Precision and Recall for Ontology Matching. In Benjamin Ashpole, Marc Ehrig, Jérôme Euzenat, and Heiner Stuckenschmidt, editors, *Proceedings of the K-CAP 2005 Workshop on Integrating Ontologies*, volume 156 of *CEUR Workshop Proceedings*, pages 25–32. CEUR-WS.org, October 2005.

[43] Agoston E. Eiben and James E. Smith. *Introduction to Evolutionary Computing*. Natural Computing Series. Springer, Berlin, Heidelberg, 2nd edition edition, 2007.

[44] Andries P. Engelbrecht. *Fundamentals of Computational Swarm Intelligence*. Wiley, 2007.

[45] Jérôme Euzenat, Alfio Ferrara, Laura Hollink, Antoine Isaac, Cliff Joslyn, Véronique Malaisé, Christian Meilicke, Andriy Nikolov,

Juan Pane, Marta Sabou, François Scharffe, Pavel Shvaiko, Vassilis Spiliopoulos, Heiner Stuckenschmidt, Ondřej Šváb-Zamazal, Vojtěch Svátek, Cássia Trojahn dos Santos, George Vouros, and Shenghui Wang. Results of the Ontology Alignment Evaluation Initiative 2009. In Pavel Shvaiko, Jérôme Euzenat, Fausto Giunchiglia, Heiner Stuckenschmidt, Natasha Noy, and Arnon Rosenthal, editors, *Proceedings of the 4th International Workshop on Ontology Matching (OM-2009)*, volume 551, pages 73–126, `http://ceur-ws.org`, October 2009. CEUR Workshop Proceedings.

[46] Jérôme Euzenat, Alfio Ferrara, Christian Meilicke, Andriy Nikolov, Juan Pane, François Scharffe, Pavel Shvaiko, Heiner Stuckenschmidt, Ondřej Šváb-Zamazal, Vojtěch Svátek, and Cássia Trojahn dos Santos. Results of the Ontology Alignment Evaluation Initiative 2010. In Pavel Shvaiko, Jérôme Euzenat, Fausto Giunchiglia, Heiner Stuckenschmidt, Ming Mao, and Isabel Cruz, editors, *Proceedings of the 5th International Workshop on Ontology Matching (OM-2010)*, volume 689, pages 85–117, `http://ceur-ws.org`, November 2010. CEUR Workshop Proceedings.

[47] Jérôme Euzenat, Alfio Ferrara, Willem Robert van Hage, Laura Hollink, Christian Meilicke, Andriy Nikolov, François Scharffe, Pavel Shvaiko, Heiner Stuckenschmidt, Ondřej Šváb-Zamazal, and Cássia Trojahn dos Santos. Final results of the Ontology Alignment Evaluation Initiative 2011. In Pavel Shvaiko, Jérôme Euzenat, Tom Heath, Christoph Quix, Ming Mao, and Isabel Cruz, editors, *Proceedings of the 6th International Workshop on Ontology Matching (OM-2011)*, volume 814, pages 85–113, `http://ceur-ws.org`, October 2011. CEUR Workshop Proceedings.

[48] Jérôme Euzenat, Antoine Isaac, Christian Meilicke, Pavel Shvaiko, Heiner Stuckenschmidt, Ondřej Šváb, Vojtěch Svátek, Willem Robert van Hage, and Mikalai Yatskevich. Results of the Ontology Alignment Evaluation Initiative 2007. In Pavel Shvaiko, Jérôme Euzenat, Fausto Giunchiglia, and Bin He, editors, *Proceedings of the 2nd International Workshop on Ontology Matching (OM-2007)*, volume 304, pages 96–132, `http://ceur-ws.org`, November 2007. CEUR Workshop Proceedings.

[49] Jérôme Euzenat, Christian Meilicke, Heiner Stuckenschmidt, Pavel Shvaiko, and Cássia Trojahn dos Santos. Ontology Alignment Eval-

uation Initiative: Six Years of Experience. In Stefano Spaccapietra, editor, *Journal of Data Semantics XV*, volume 6720 of *Lecture Notes in Computer Science*, pages 158–192. Springer, Berlin, Heidelberg, 2011.

[50] Jérôme Euzenat and Pavel Shvaiko. *Ontology Matching*. Springer, Berlin, Heidelberg, 2007.

[51] Jérôme Euzenat, Heiner Stuckenschmidt, and Mikalai Yatskevich. Introduction to the Ontology Alignment Evaluation 2005. In Benjamin Ashpole, Marc Ehrig, Jérôme Euzenat, and Heiner Stuckenschmidt, editors, *Proceedings of the K-CAP 2005 Workshop on Integrating Ontologies*, volume 156, `http://ceur-ws.org`, October 2005.

[52] Michael J. Fogel, Lawrence J. Owens, and Alvin J. Walsh. *Artificial Intelligence through Simulated Evolution*. Wiley, Chichester, WS, UK, 1966.

[53] Fred Freitas, Stefan Schulz, and Eduardo Moraes. Survey of current terminologies and ontologies in biology and medicine. *Electronic Journal of Communication Information & Innovation in Health*, 3(1):7–18, March 2009.

[54] Michael Gelfond and Vladimir Lifschitz. Classical Negation in Logic Programs and Disjunctive Databases. *New Generation Computing*, 9(3–4):365–386, 1991.

[55] Fausto Giunchiglia and Pavel Shvaiko. Semantic Matching. *The Knowledge Engineering Review*, 18(3):265–280, May 2004.

[56] Fausto Giunchiglia, Pavel Shvaiko, and Mikalai Yatskevich. S-Match: An Algorithm and an Implementation of Semantic Matching. In Christoph Bussler, John Davies, Dieter Fensel, and Rudi Studer, editors, *Proceedings of the 1st European Semantic Web Symposium*, volume 3053 of *Lecture Notes in Computer Science*, pages 61–75, Berlin, Heidelberg, May 2004. Springer.

[57] Michael Granitzer, Vedran Sabol, Kow Weng Onn, Dickson Lukose, and Klaus Tochtermann. Ontology Alignment—A Survey with Focus on Visually Supported Semi-Automatic Techniques. *Future Internet*, 2:238–258, August 2010.

[58] John Grefenstette. Rank-based selection. In Thomas Bäck, David B. Fogel, and Zbigniew Michalewicz, editors, *Evolutionary Computation 1: Basic Algorithms and Operators*, chapter 25, pages 187–194. Taylor & Francis, New York, NY, USA, 2000.

[59] Thomas R. Gruber. Toward Principles for the Design of Ontologies Used for Knowledge Sharing. In Nicola Guarino and Roberto Poli, editors, *Formal Ontology in Conceptual Analysis and Knowledge Representation*, Dordrecht, The Netherlands, 1993. Kluwer Academic Publishers.

[60] Christophe Guéret, Eyal Oren, Stefan Schlobach, and Martijn Schut. An Evolutionary Perspective on Approximate RDF Query Answering. In Sergio Greco and Thomas Lukasiewicz, editors, *Proceedings of the 2nd International Conference on Scalable Uncertainty Management*, volume 5291 of *Lecture Notes in Computer Science*, pages 215–228, Berlin, Heidelberg, October 2008. Springer.

[61] Ramanathan V. Guha and Dan Brickley. RDF Vocabulary Description Language 1.0: RDF Schema. W3C recommendation, W3C, February 2004.

[62] Tim Hendtlass, Irene Moser, and Marcus Randall. Dynamic Problems and Nature Inspired Meta-heuristics. In Andrew Lewis, Sanaz Mostaghim, and Marcus Randall, editors, *Biologically-Inspired Optimisation Methods*, volume 210 of *Studies in Computational Intelligence*, pages 79–109. Springer, Berlin, Heidelberg, 2009.

[63] Jan Hettenhausen. Interactive Multi-Objective Particle Swarm Optimisation with Heatmap Visualisation based User Interface. Master's thesis, School of Information and Communication Technology, Griffith University, Nathan campus, Brisbane, QLD, 4111, Australia, 2007.

[64] David Hilley. Cloud Computing: A Taxonomy of Platform and Infrastructure-level Offerings. Technical Report GIT-CERCS-09-13, College of Computing, Georgia Institute of Technology, Atlanta, GA, United States, April 2009.

[65] Pascal Hitzler, Markus Krötzsch, and Sebastian Rudolph. *Foundations of Semantic Web Technologies*. Chapman & Hall/CRC textbooks in computing. CRC Press, Boca Raton, FL, USA, 2010.

[66] John H. Holland. *Adaptation in Natural and Artificial Systems*. University of Michigan Press, Ann Arbor, MI, United States, 1975.

[67] Matthew Horridge and Sean Bechhofer. The OWL API: A Java API for Working with OWL 2 Ontologies. In Rinke Hoekstra and Peter F. Patel-Schneider, editors, *Proceedings of the 6th International Workshop on OWL: Experiences and Directions (OWLED 2009)*, volume 529, http://ceur-ws.org, 2009. CEUR Workshop Proceedings.

[68] Ian Horrocks, Oliver Kutz, and Ulrike Sattler. The Even More Irresistible $\mathcal{SROIQ}$. In Patrick Doherty, John Mylopoulos, and Christopher A. Welty, editors, *Proceedings of the 10th International Conference on Principles of Knowledge Representation and Reasoning*, pages 57–67, Menlo Park, California, USA, June 2006. AAAI Press.

[69] Yves R. Jean-Mary, E. Patrick Shironoshita, and Mansur R. Kabuka. Ontology Matching with Semantic Verification. *Web Semantics*, 7(3):235–251, September 2009.

[70] Qiu Ji, Peter Haase, and Guilin Qi. Combination of Similarity Measures in Ontology Matching using the OWA Operator. In *Proceedings of the 12th International Conference on Information Processing and Management of Uncertainty in Knowledge-Base Systems (IPMU'08)*, 2008.

[71] Cliff A. Joslyn, Patrick Paulson, and Amanda White. Measuring the Structural Preservation of Semantic Hierarchy Alignments. In Pavel Shvaiko, Jérôme Euzenat, Fausto Giunchiglia, Heiner Stuckenschmidt, Natasha Noy, and Arnon Rosenthal, editors, *Proceedings of the 4th International Workshop on Ontology Matching*, volume 551, pages 61–72, http://ceur-ws.org, October 2009. CEUR Workshop Proceedings.

[72] Aditya A. Kalyanpur. *Debugging and Repair of OWL Ontologies*. PhD thesis, Faculty of the Graduate School, University of Maryland, College Park, MD, USA, 2006.

[73] Stuart Kauffman and Simon Levin. Towards a General Theory of Adaptive Walks on Rugged Landscapes. *Journal of Theoretical Biology*, 128(1):11–45, September 1987.

[74] Stuart A. Kauffman and Edward D. Weinberger. The NK Model of Rugged Fitness Landscapes And Its Application to Maturation of the Immune Response. *Journal of Theoretical Biology*, 141(2):211–245, November 1989.

[75] James Kennedy and Russel C. Eberhart. *Swarm Intelligence*. Morgan Kaufmann, April 2001.

[76] James Kennedy and Russell C. Eberhart. Particle Swarm Optimization. In *Proceedings of IEEE International Conference on Neural Networks*, volume 4, pages 1942–1948, Washington, DC, USA, November 1995. IEEE Computer Society.

[77] James Kennedy and Russell C. Eberhart. A Discrete Binary Version of the Particle Swarm Algorithm. In *Proceedings of the IEEE International Conference on Systems, Man, and Cybernetics*, volume 5, pages 4104–4108, Washington, DC, USA, October 1997. IEEE Computer Society.

[78] Tom Killalea. Building Scalable Web Services. *ACM Queue*, 6(6):10–13, October 2008.

[79] John R. Koza. *Genetic Programming: On the Programming of Computers by Means of Natural Selection*. MIT Press, Cambridge, MA, United States, 1992.

[80] Harold W. Kuhn. The Hungarian Method for the Assignment Problem. *Naval Research Logistics Quarterly*, 2(1–2):83–97, March 1955.

[81] Alexander Lenk, Markus Klems, Jens Nimis, Stefan Tai, and Thomas Sandholm. What's inside the Cloud? An architectural map of the Cloud landscape. In *Proceedings of the ICSE Workshop on Software Engineering Challenges of Cloud Computing (CLOUD)*, pages 23–31, Washington, DC, USA, 2009. IEEE Computer Society.

[82] Vladimir Levenshtein. Binary codes capable of correcting deletions, insertions, and reversals. *Soviet Physics Doklady*, 10(8):707–710, February 1966.

[83] Andrew Lewis, David Abramson, and Tom Peachey. An Evolutionary Programming Algorithm for Automatic Engineering Design. In Roman Wyrzykowski, Jack Dongarra, Marcin Paprzycki, and Jerzy

Waśniewski, editors, *Proceedings of the 5th International Conference on Parallel Processing and Applied Mathematics*, volume 3019 of *Lecture Notes in Computer Science*, pages 586–594, Berlin, Heidelberg, September 2003. Springer.

[84] Andrew Lewis, Sanaz Mostaghim, and Ian Scriven. Asynchronous Multi-Objective Optimisation in Unreliable Distributed Environments. In Andrew Lewis, Sanaz Mostaghim, and Marcus Randall, editors, *Biologically-Inspired Optimisation Methods*, volume 210 of *Studies in Computational Intelligence*, pages 51–78. Springer, Berlin, Heidelberg, 2009.

[85] Juanzi Li, Jie Tang, Yi Li, and Qiong Luo. RiMOM: A Dynamic Multistrategy Ontology Alignment Framework. *IEEE Transactions on Knowledge and Data Engineering*, 21(8):1218–1232, August 2009.

[86] Jorge Martinez-Gil, Enrique Alba, and José F. Aldana Montes. Optimizing Ontology Alignments by Using Genetic Algorithms. In Christophe Guéret, Pascal Hitzler, and Stefan Schlobach, editors, *Nature inspired Reasoning for the Semantic Web (NatuReS)*, volume 419, http://ceur-ws.org, October 2008. CEUR Workshop Proceedings.

[87] Jorge Martinez-Gil and José F. Aldana-Montes. Evaluation of two Heuristic Approaches to Solve the Ontology Meta-Matching Problem. *Knowledge and Information Systems*, 26(2):225–247, 2011.

[88] Christian Meilicke and Heiner Stuckenschmidt. Repairing Ontology Mappings. In Robert C. Holte and Adele Howe, editors, *Proceedings of the 22nd AAAI Conference on Artificial Intelligence*, pages 1408–1413, Menlo Park, CA, USA, July 2007. AAAI Press.

[89] Christian Meilicke, Cássia Trojahn dos Santos, Jakob Huber, and Jérôme Euzenat. Integrating Ontology Matching Tools into the SEALS Platform. Tutorial, SEALS Project, February 2012.

[90] Sergey Melnik, Hector Garcia-Molina, and Erhard Rahm. Similarity flooding: a versatile graph matching algorithm and its application to schema matching. In Dimitrios Georgakopoulos, Rakesh Agrawal, and Klaus Dittrich, editors, *Proceedings of the 18th International Conference on Data Engineering*, pages 117–128, Washington, DC, USA, February 2002. IEEE Computer Society.

[91] Peter Mika and Giovanni Tummarello. Web Semantics in the Clouds. *IEEE Intelligent Systems*, 23(5):82–87, September 2008.

[92] Boris Motik, Bijan Parsia, and Peter F. Patel-Schneider. OWL 2 Web Ontology Language Structural Specification and Functional-Style Syntax. W3C recommendation, W3C, October 2009.

[93] James R. Munkres. Algorithms for the Assignment and Transportation Problems. *Journal of the Society for Industrial and Applied Mathematics*, 5(1):32–38, March 1957.

[94] Andre Newman, Yuan-Fang Li, and Jane Hunter. Scalable Semantics – the Silver Lining of Cloud Computing. In *Proceedings of the IEEE Fourth International Conference on eScience*, pages 111–118, Washington, DC, USA, December 2008. IEEE Computer Society.

[95] Mathias Niepert, Christian Meilicke, and Heiner Stuckenschmidt. A Probabilistic-Logical Framework for Ontology Matching. In Maria Fox and David Poole, editors, *Proceedings of the 24th AAAI Conference on Artificial Intelligence*, pages 1413–1418, Menlo Park, California, USA, July 2010. AAAI Press.

[96] Natalya F. Noy, Nigam H. Shah, Patricia L. Whetzel, Benjamin Dai, Michael Dorf, Nicholas Griffith, Clement Jonquet, Daniel L. Rubin, Margaret-Anne Storey, Christopher G. Chute, and Mark A. Musen. Bioportal: ontologies and integrated data resources at the click of a mouse. *Nucleic Acids Research*, 37:170–173, May 2009.

[97] Eyal Oren, Christophe Guéret, and Stefan Schlobach. Anytime Query Answering in RDF through Evolutionary Algorithms. In Amit P. Sheth, Steffen Staab, Mike Dean, Massimo Paolucci, Diana Maynard, Timothy W. Finin, and Krishnaprasad Thirunarayan, editors, *Proceedings of the 7th International Semantic Web Conference*, volume 5318 of *Lecture Notes in Computer Science*, pages 98–113, Berlin, Heidelberg, October 2008. Springer.

[98] Gary Pampará, Andries P. Engelbrecht, and Nelis Franken. Binary Differential Evolution. In *Proceedings of the IEEE Congress on Evolutionary Computation*, pages 1873–1879, Washington, DC, USA, July 2006. IEEE Computer Society.

[99] Peter F. Patel-Schneider, Boris Motik, and Bernardo Cuenca Grau. OWL 2 Web Ontology Language Direct Semantics. W3C recommendation, W3C, October 2009.

[100] Heiko Paulheim. Skalierbarkeit von Ontology-Matching-Verfahren. Master's thesis, Technische Universität Darmstadt, February 2008.

[101] Valentina Presutti, Eva Blomqvist, Enrico Daga, and Aldo Gangemi. Pattern-based ontology design. In Mari Carmen Suárez-Figueroa, Asunción Gómez-Pérez, Enrico Motta, and Aldo Gangemi, editors, *Ontology Engineering in a Networked World*, pages 35–64. Springer, Berlin, Heidelberg, 2012.

[102] Guilin Qi and Anthony Hunter. Measuring Incoherence in Description Logic-based Ontologies. In Karl Aberer, Philippe Cudré-Mauroux, Key-Sun Choi, Natasha Noy, Dean Allemang, Kyung-Il Lee, Lyndon Nixon, Jennifer Golbeck, Peter Mika, Diana Maynard, Riichiro Mizoguchi, and Guus Schreiber, editors, *Proceedings of the 6th International Semantic Web Conference*, volume 4825 of *Lecture Notes in Computer Science*, pages 381–394, Berlin, Heidelberg, November 2007. Springer.

[103] Yuzuhong Qu, Wei Hu, and Gong Cheng. Constructing Virtual Documents for Ontology Matching. In Les Carr, David De Roure, Arun Iyengar, Carole A. Goble, and Michael Dahlin, editors, *Proceedings of the 15th International World Wide Web Conference*, pages 23–31, New York, USA, May 2006. ACM.

[104] Erhard Rahm. Towards Large-Scale Schema and Ontology Matching. In Zohra Bellahsene, Angela Bonifati, and Erhard Rahm, editors, *Schema Matching and Mapping*, Data-Centric Systems and Applications, chapter 1, pages 3–28. Springer, Berlin, Heidelberg, 2011.

[105] Marcus Randall. Enhancements to Extremal Optimisation for Generalised Assignment. In Marcus Randall, Hussein A. Abbass, and Janet Wiles, editors, *Progress in Artificial Life*, volume 4828 of *Lecture Notes in Artificial Intelligence*, pages 369–380. Springer, Berlin, Heidelberg, 2007.

[106] Marcus Randall, Tim Hendtlass, and Andrew Lewis. Extremal Optimisation for Assignment Type Problems. In Andrew Lewis, Sanaz

Mostaghim, and Marcus Randall, editors, *Biologically-Inspired Optimisation Methods*, volume 210 of *Studies in Computational Intelligence*, pages 139–164. Springer, Berlin, Heidelberg, 2009.

[107] Sanjaya Ratnayake, Ruvindee Rupasinghe, Ranatunga Anuruddha, Shalinda Adikari, Sajayahan de Zoysa, Kamala Tennakoon, and Asoka Karunananda. Using swarm Intelligence to perceive the Semantic Web. In *Proceedings of the 4th International Conference on Information and Automation for Sustainability*, pages 91–96, Washington, DC, USA, December 2008. IEEE Computer Society.

[108] Ingo Rechenberg. *Evolutionsstrategie: Optimierung technischer Systeme nach Prinzipien der biologischen Evolution*. Frommann Holzboog, Stuttgart, Germany, 1973.

[109] Sebastian Riedel. Improving the Accuracy and Efficiency of MAP Inference for Markov Logic. In David A. McAllester and Petri Myllymki, editors, *Proceedings of the 24th Conference on Uncertainty in Artificial Intelligence*, pages 468–475. AUAI Press, July 2008.

[110] Dominique Ritze and Heiko Paulheim. Towards an Automatic Parameterization of Ontology Matching Tools based on Example Mappings. In Pavel Shvaiko, Jérôme Euzenat, Tom Heath, Christoph Quix, Ming Mao, and Isabel Cruz, editors, *Proceedings of the 6th International Workshop on Ontology Matching*, volume 814, pages 37–48, http://ceur-ws.org, October 2011. CEUR Workshop Proceedings.

[111] Maria-Elena Roşoiu, Cássia Trojahn dos Santos, and Jérôme Euzenat. Ontology Matching Benchmarks: Generation and Evaluation. In Pavel Shvaiko, Jérôme Euzenat, Tom Heath, Christoph Quix, Ming Mao, and Isabel Cruz, editors, *Proceedings of the 6th International Workshop on Ontology Matching (OM-2011)*, volume 814, pages 73–84, http://ceur-ws.org, October 2011. CEUR Workshop Proceedings.

[112] Sebastian Rudolph, Tuvshintur Tserendorj, and Pascal Hitzler. What is Approximate Reasoning? In Diego Calvanese and Georg Lausen, editors, *Proceedings of the 2nd International Conference on Web Reasoning and Rule Systems (RR2008)*, volume 5341 of *Lecture Notes in Computer Science*, pages 150–164, Berlin, Heidelberg, November 2008. Springer.

[113] Gerard M. Salton, Andrew K. C. Wong, and Chung-Shu Yang. A Vector Space Model for Automatic Indexing. *Communications of the ACM*, 18(11):613–620, November 1975.

[114] Frederik C. Schadd and Nico Roos. MaasMatch results for OAEI 2011. In Pavel Shvaiko, Jérôme Euzenat, Tom Heath, Christoph Quix, Ming Mao, and Isabel Cruz, editors, *Proceedings of the 6th International Workshop on Ontology Matching (OM-2011)*, volume 814, pages 171–178, http://ceur-ws.org, October 2011. CEUR Workshop Proceedings.

[115] Hans-Paul Schwefel. *Evolutionsstrategie und numerische Optimierung*. PhD thesis, Technical University of Berlin, Department of Process Engineering, Berlin, Germany, 1975.

[116] Yuhui Shi and Russell C. Eberhart. A Modified Particle Swarm Optimizer. In *Proceedings of IEEE International Conference on Evolutionary Computation*, pages 69–73, Washington, DC, USA, May 1998. IEEE Computer Society.

[117] Barry Smith, Michael Ashburner, Cornelius Rosse, Jonathan Bard, William Bug, Werner Ceusters, Louis J. Goldberg, Karen Eilbeck, Amelia Ireland, Christopher J. Mungall, Neocles Leontis, Philippe Rocca-Serra, Alan Ruttenberg, Susanna-Assunta Sansone, Richard H. Scheuermann, Nigam Shah, Patricia L. Whetzel, and Suzanna Lewis. The OBO Foundry: Coordinated Evolution of Ontologies to Support Biomedical Data Integration. *Nature Biotechnology*, 25(11):1251–1255, November 2007.

[118] William M. Spears. Crossover or Mutation? In *Foundations of Genetic Algorithms 2*, pages 221–238. Morgan Kaufmann, 1993.

[119] Raffael Stein and Valentin Zacharias. RDF On Cloud Number Nine. In Stefano Ceri, Emanuele Della Valle, Jim Hendler, and Zhisheng Huang, editors, *Proceedings of the 4th Workshop on New Forms of Reasoning for the Semantic Web: Scalable & Dynamic*, pages 11–23, http://ceur-ws.org, May 2010. CEUR Workshop Proceedings.

[120] Giorgos Stoilos, Giorgos Stamou, and Stefanos Kollias. A String Metric For Ontology Alignment. In Yolanda Gil, Enrico Motta, V. Richard Benjamins, and Mark A. Musen, editors, *Proceedings of the 4rd International Semantic Web Conference (ISWC)*, volume 3729

of *Lecture Notes in Computer Science*, pages 624–637, Berlin, Heidelberg, November 2005. Springer.

[121] Rainer Storn and Kenneth Price. Differential Evolution – A Simple and Efficient Heuristic for Global Optimization over Continuous Spaces. *Journal of Global Optimization*, 11(4):341–359, December 1997.

[122] Rudi Studer, V. Richard Benjamins, and Dieter Fensel. Knowledge Engineering: Principles and Methods. *Data & Knowledge Engineering*, 25(1–2):161–197, March 1998.

[123] Jie Tang, Juanzi Li, Bangyong Liang, Xiaotong Hunag, Yi Li, and Kehong Wang. Using Bayesian decision for ontology mapping. *Web Semantics: Science, Services and Agents on the World Wide Web*, 4(4):243–262, December 2006.

[124] Jie Tang, Bang-Yong Liang, Juanzi Li, and Kehong Wang. Risk Minimization Based Ontology Mapping. In Chi-Hung Chi and Kwok-Yan Lam, editors, *Proceedings of the Advanced Workshop on Content Computing*, volume 3309 of *Lecture Notes in Computer Science*, pages 469–480, Berlin, Heidelberg, November 2004. Springer.

[125] The Gene Ontology Consortium. The Gene Ontology project in 2008. *Nucleic Acids Research*, 36(1):440–444, January 2008.

[126] René Thomsen. Flexible ligand docking using evolutionary algorithms: investigating the effects of variation operators and local search hybrids. *Biosystems*, 72(1–2):57–73, November 2003.

[127] Hoai-Viet To, Ryutaro Ichise, and Hoai-Bac Le. An Adaptive Machine Learning Framework with User Interaction for Ontology Matching. In Biplav Srivastava, Ullas Nambiar, and Craig Knoblock, editors, *Proceedings of IJCAI Workshop on Information Integration on the Web*, pages 35–40, Pasadena, CA, USA, July 2009.

[128] Jacopo Urbani, Spyros Kotoulas, Jason Maassen, Frank van Harmelen, and Henri Bal. OWL Reasoning with WebPIE: Calculating the Closure of 100 Billion Triples. In Lora Aroyo, Grigoris Antoniou, Eero Hyvnen, Annette ten Teije, Heiner Stuckenschmidt, Liliana Cabral, and Tania Tudorache, editors, *Proceedings of the 7th Extended Semantic Web Conference (ESWC)*, volume 6088 of *Lecture*

Notes in Computer Science, pages 213–227, Berlin, Heidelberg, May 2010. Springer.

[129] Jacopo Urbani, Spyros Kotoulas, Eyal Oren, and Frank van Harmelen. Scalable Distributed Reasoning using MapReduce. In Abraham Bernstein, David R. Karger, Tom Heath, Lee Feigenbaum, Diana Maynard, Enrico Motta, and Krishnaprasad Thirunarayan, editors, *Proceedings of the 8th International Semantic Web Conference (ISWC)*, volume 5823 of *Lecture Notes in Computer Science*, pages 634–649, Berlin, Heidelberg, October 2009. Springer.

[130] Junli Wang, Zhijun Ding, and Changjun Jiang. GAOM: Genetic Algorithm Based Ontology Matching. In *Proceedings of the IEEE Asia-Pacific Conference on Services Computing (APSCC)*, pages 617–620, Washington, DC, USA, 2006. IEEE Computer Society.

[131] Shenghui Wang, Gwenn Englebienney, Christophe Guéret, Stefan Schlobach, Antoine Isaac, and Martijn Schut. Similarity Features, and their Role in Concept Alignment Learning. In Manuela Popescu and Darin L. Stewart, editors, *Proceedings of the 4th International Conference on Advances in Semantic Processing*, pages 1–6, `http://www.xpertps.com`, October 2010. Xpert Publishing Services.

[132] Darrell Whitley. Recombination – permutations. In Thomas Bäck, David B. Fogel, and Zbigniew Michalewicz, editors, *Evolutionary Computation 1: Basic Algorithms and Operators*, chapter 33.3, pages 274–284. Taylor & Francis, New York, NY, USA, 2000.

[133] William E. Winkler. The State of Record Linkage and Current Research Problems. Technical report, Statistical Research Division, Bureau of the Census, Washington, DC, USA, 1999.

[134] Xiao Zhang, Qian Zhong, Juanzi Li, and Jie Tang. RiMOM results for OAEI 2008. In Pavel Shvaiko, Jérôme Euzenat, Fausto Giunchiglia, and Heiner Stuckenschmidt, editors, *Proceedings of the 3rd International Workshop on Ontology Matching*, volume 431, `http://ceur-ws.org`, 2008. CEUR Workshop Proceedings.

A. Ontology Alignment Evaluation Initiative *benchmarks*

The following figures illustrate the detailed results of the MapPSO prototype in the *benchmarks* track of the Ontology Alignment Evaluation Initiative (OAEI) campaigns from 2008 till 2010. Precision and recall evaluations for each year are shown in separate plots. Each plot shows the score of MapPSO for each test case in the *benchmarks* track. Both classical precision and recall, as well as symmetric precision and recall (cf. Section 7.1) are shown[1].

In the *benchmarks* data set, a single ontology is first aligned with itself (test case 101) and systematically to alterations of itself. Test cases 30x) are alignments of the original *benchmarks* ontology with real-world ontologies found on the Web. For details about the structure of the data set and the details of the different alterations, see

```
http://oaei.ontologymatching.org/2008/benchmarks/index.html
http://oaei.ontologymatching.org/2009/benchmarks/index.html
http://oaei.ontologymatching.org/2010/benchmarks/index.html
```

---

[1]Since the symmetric measure is always greater or equal to the classical measure by definition, the bars indicating the classical measure is drawn in front of the symmetric measure.

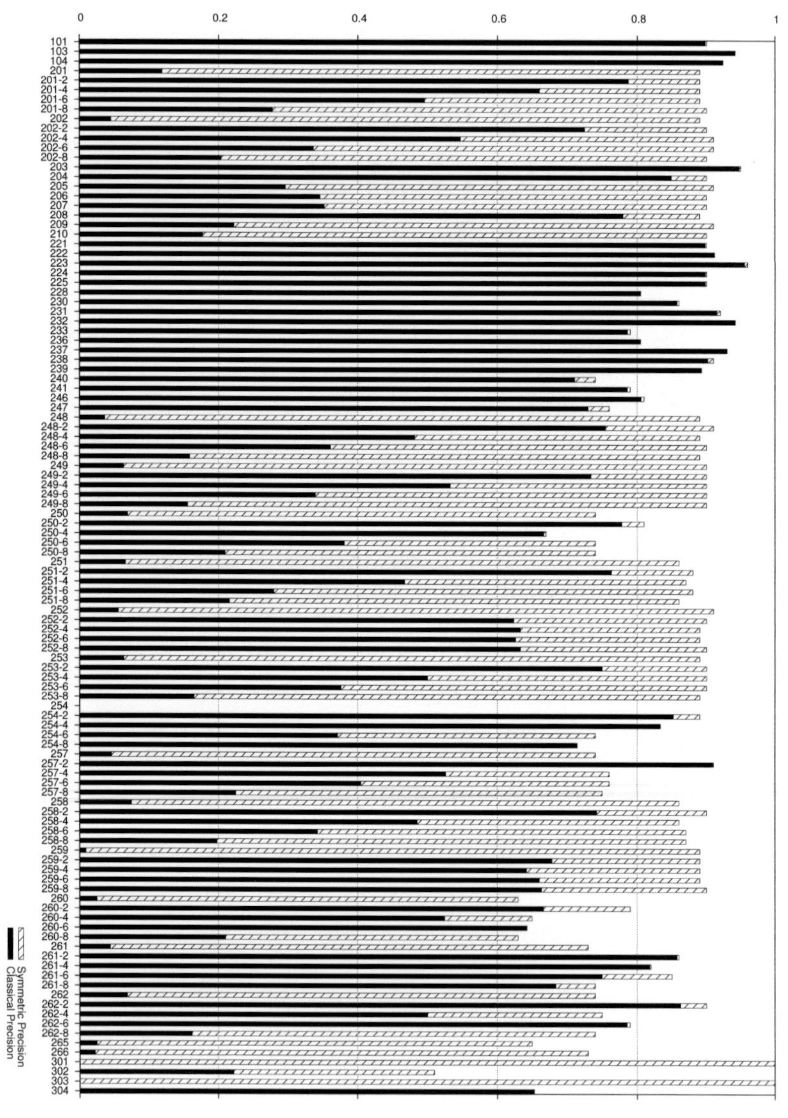

Figure A.1.: Classical and symmetric precision for the OAEI 2008 *benchmarks* track.

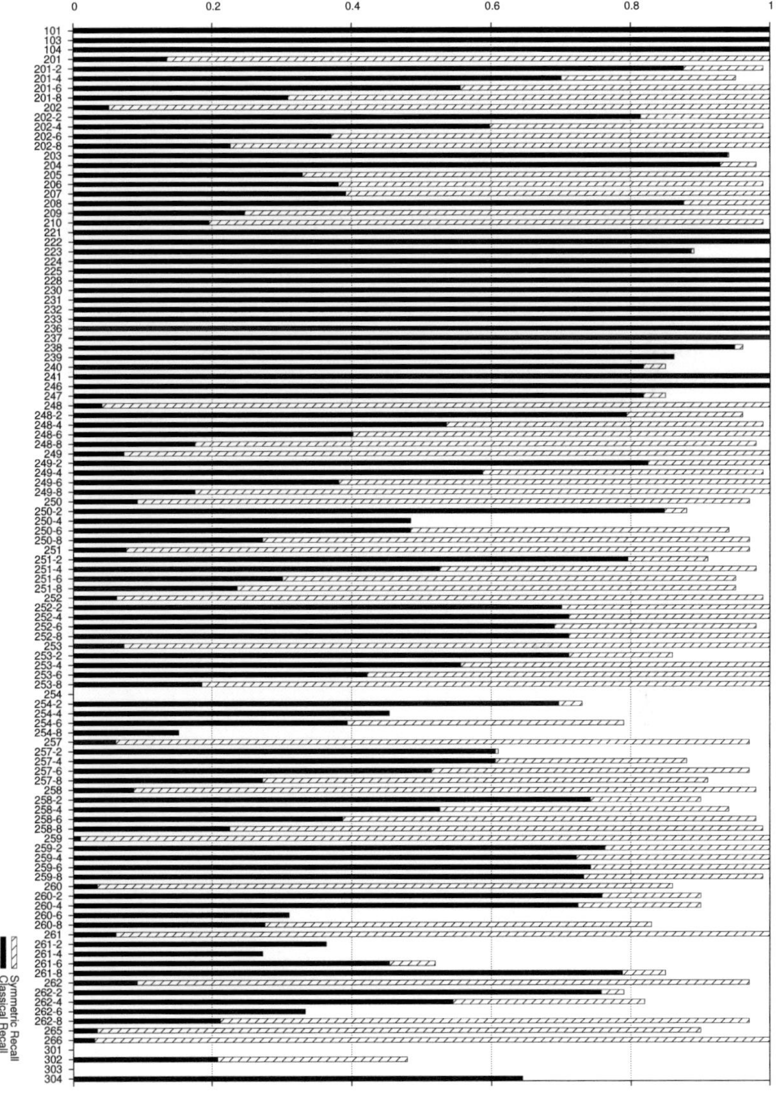

Figure A.2.: Classical and symmetric recall for the OAEI 2008 *benchmarks* track.

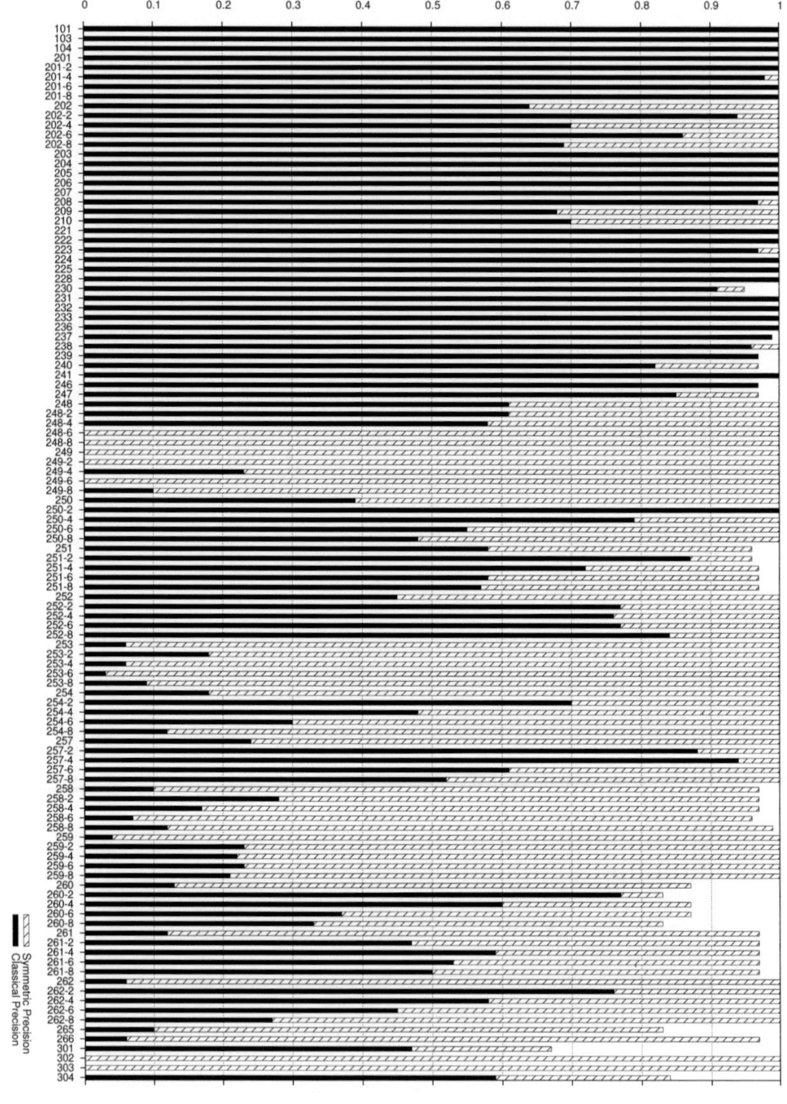

Figure A.3.: Classical and symmetric precision for the OAEI 2009 *benchmarks* track.

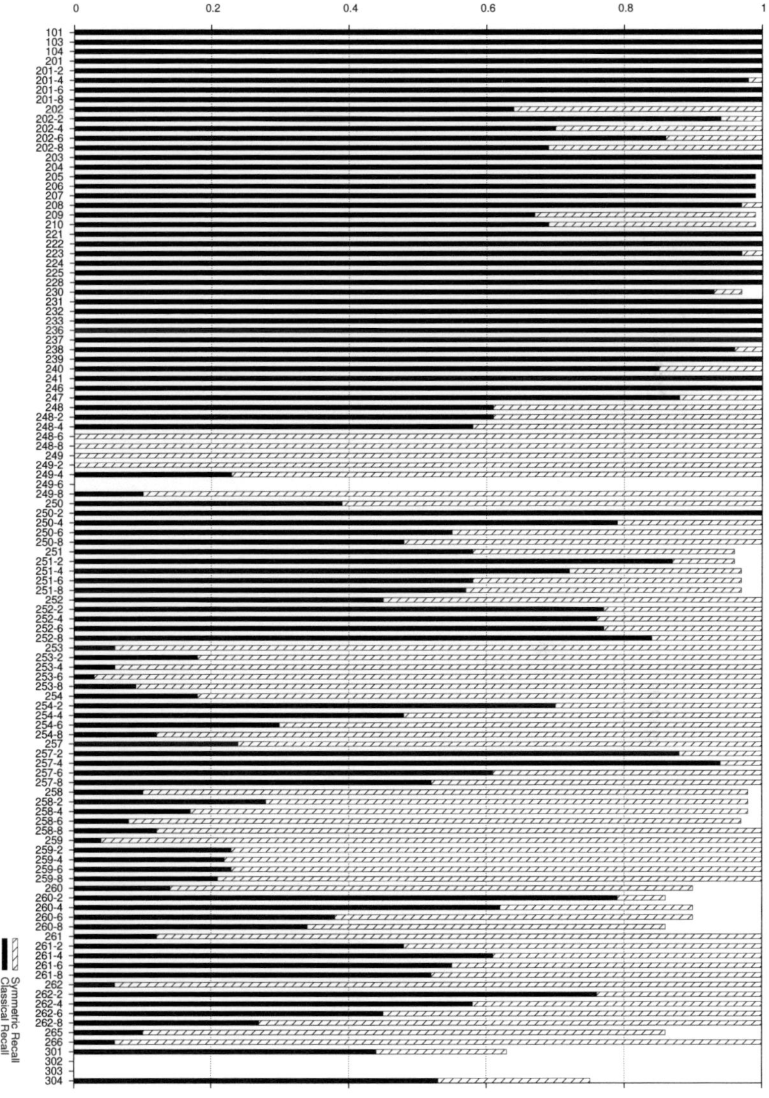

Figure A.4.: Classical and symmetric recall for the OAEI 2009 *benchmarks* track.

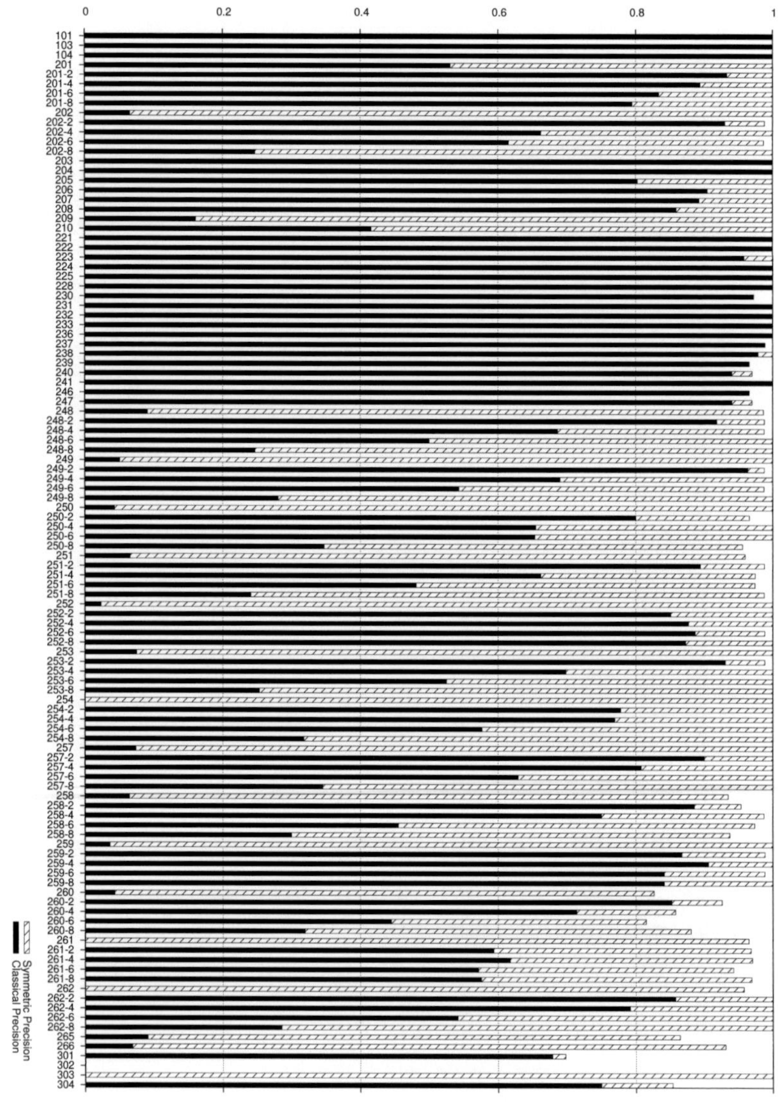

Figure A.5.: Classical and symmetric precision for the OAEI 2010 *benchmarks* track.

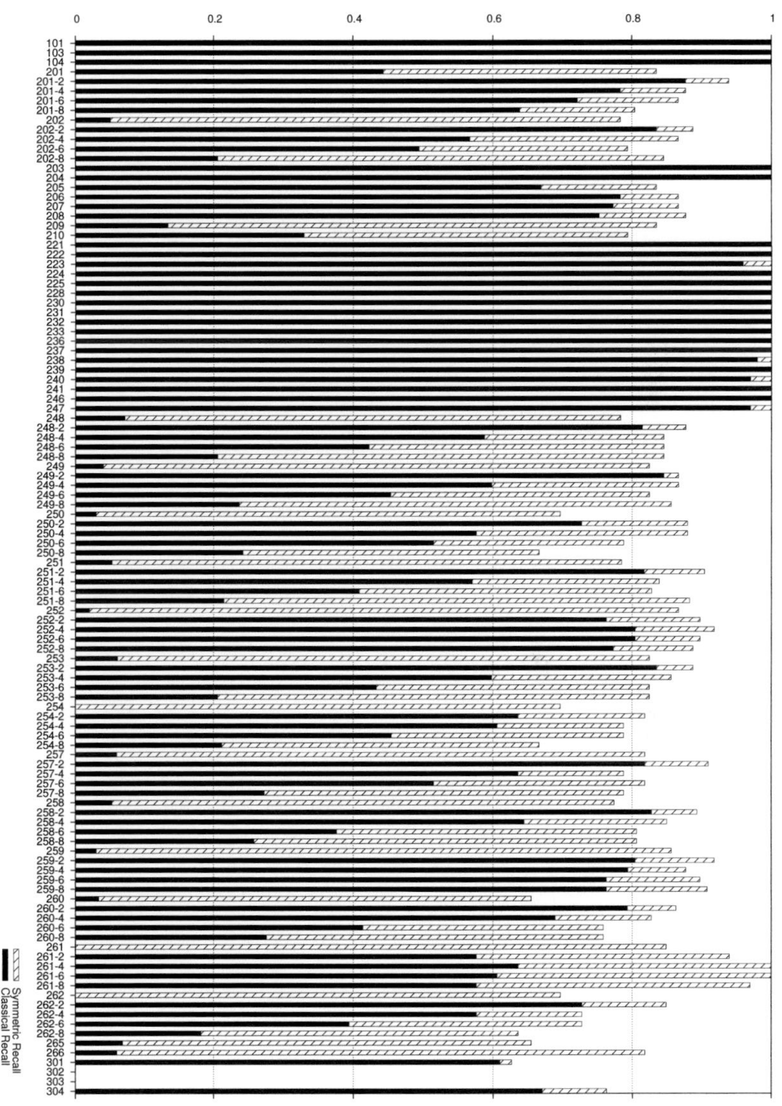

Figure A.6.: Classical and symmetric recall for the OAEI 2010 *benchmarks* track.

B. Parameter Configurations used in Evaluations

The following tables list the parameter configurations for MapEVO and MapPSO used in the Ontology Alignment Evaluation Initiative (OAEI) in the years 2011 and 2012. The evaluations were carried out in the *SEALS* platform, which imposes several restrictions, *e.g.* on the population size to be chosen.

The evaluation modalities required a single configuration to be used for all tracks in the campaign. Since no sophisticated self-adaptation procedure was developed for the presented prototypes, these configurations were chosen rather conservatively. However, for the OAEI 2012, a very simple self-adaptation was implemented in order to adjust the number of iterations to be performed depending on the size of the larger ontology. This is required to ensure the number of iterations to be large enough to allow for convergence, and small enough to safe time after convergence.

Since all evaluations are carried out on a single machine in the *SEALS* platform, a relatively small population is used in all settings. While in 2011 classes, object properties, and data properties correspondences were considered, in 2012 only class correspondences were considered, since there is no need for property correspondences in the *anatomy* and *large biomedical ontologies* tracks. However, in this setting the recall scores in the *benchmarks* and *conference* tracks are expected to drop, since properties are part of the reference alignments there. The algorithm internal settings were kept stable, apart from the ϱ parameter which was introduced after the 2011 campaign[1].

Regarding the objective function, a very basic setting was used in 2011 with only alignment size and correspondence contribution being used on the alignment level. The correspondence contribution is the mean value of correspondence confidences (cf. Equation (4.43)). Each correspondence confidence is the weighted average of a very basic collection of correspon-

---

[1]The equivalent of not incorporating the ϱ factor is equal to setting $\varrho = 1$.

dence evaluators. In 2012, a wider collection of evaluators were used, incorporating label and virtual document similarity on the correspondence level, as well as structural metrics on both the alignment and the correspondence level.

Parameter	Value	Description
i_{max}	200	number of iterations
$\lvert I \rvert$	10	population size
i_{sel}	50	number of selections (cf. Algorithm 5.2)
ζ	0.25	selection ratio (cf. Algorithm 5.2)
ϱ	1.0	swap limitation (cf. Equation (5.2))
a in p_{change}	0.5	cf. Equations (5.7) and (5.8)
a in p_{setN}	0.01	correspondence removal probability lower bound (cf. Equation (5.6))
b in p_{setN}	0.1	correspondence removal probability upper bound (cf. Equation (5.6))
a in p_{setV}	0.02	correspondence adding probability lower bound (cf. Equation (5.5))
b in p_{setV}	0.2	correspondence adding probability upper bound (cf. Equation (5.5))
objective function		$$F(A) \;=\; \Gamma_{\text{weightAvg}}((H_{\text{size}}(A),\, H_{\text{corrContrib}}(A)),\, (0.6,\, 0.4))$$ where for all $C \in A$ $$\iota(C) \;=\; \Gamma_{\text{weightAvg}}((h_{\text{lexIDSim}}(C),\, h_{\text{textIDSim}}(C),\, h^{A}_{\text{hierarchy}}(C),\, h^{A}_{\text{propDRClass}}(C)),\, (.25,\, .25,\, .25,\, .25))$$ which is also used as support heuristic.

Table B.1.: Configuration parameters of MapEVO for OAEI 2011.

Parameter	Value	Description
i_{max}	1000	number of iterations
$\|I\|$	20	population size
enable classes	`true`	
enable object properties	`true`	
enable data properties	`true`	
enable individuals	`false`	
κ	0.6	keep-set threshold (cf. Equation (5.19))
σ	0.1	safe-set threshold (cf. Equation (5.20))
topology	`StarTopology`	social network structure (cf. Table 6.5)
comm. strategy	synchronous	cf. Table 6.5
β	0.1	proportional likelihood increment (personal best)
γ	0.05	proportional likelihood increment (local best)
λ	0.3	size change probability factor (cf. Equation (5.24))
w_{inc} start value	2.5	cf. Equation (5.25)
w_{inc} end value	1.5	cf. Equation (5.25)
w_{dec} start value	1.5	cf. Equation (5.25)
w_{dec} end value	1.1	cf. Equation (5.25)
objective function		

$$F(A) = \Gamma_{\text{weightAvg}}((H_{\text{size}}(A),\, H_{\text{corrContrib}}(A)),$$
$$(0.6,\, 0.4))$$

where for all $C \in A$

$$\iota(C) = \Gamma_{\text{weightAvg}}((h_{\text{lexIDSim}}(C),$$
$$h_{\text{textIDSim}}(C),$$
$$h_{\text{hierarchy}}^{A}(C),$$
$$h_{\text{propDRClass}}^{A}(C)),$$
$$(.25,\, .25,\, .25,\, .25))$$

which is also used as support heuristic.

Table B.2.: Configuration parameters of MapPSO for OAEI 2011.

Parameter	Value	Description
i_{max}	$100+$ $\max\{\sharp\mathcal{O}_1, \sharp\mathcal{O}_2\}$	number of iterations
$|I|$	8	population size
i_{sel}	4	number of selections (cf. Algorithm 5.2)
ζ	0.25	selection ratio (cf. Algorithm 5.2)
ϱ	0.3	swap limitation (cf. Equation (5.2))
a in p_{change}	0.5	cf. Equations (5.7) and (5.8)
a in p_{setN}	0.01	correspondence removal probability lower bound (cf. Equation (5.6))
b in p_{setN}	0.1	correspondence removal probability upper bound (cf. Equation (5.6))
a in p_{setV}	0.02	correspondence adding probability lower bound (cf. Equation (5.5))
b in p_{setV}	0.2	correspondence adding probability upper bound (cf. Equation (5.5))
objective function	$$\begin{aligned} F(A) \;=\; & \Gamma_{\text{weightAvg}}((H_{\text{size}}(A), H_{\text{corrContrib}}(A), \\ & H_{\text{crissCross}}(A), H_{\text{structPreserv}}(A)), \\ & (0.1,\, 0.3,\, 0.1,\, 0.1)) \end{aligned}$$ where for all $C \in A$ $$\begin{aligned} \iota(C) \;=\; & \Gamma_{\text{weightAvg}}((h_{\text{lexIDSim}}(C), \\ & h_{\text{lexLabelSim}}(C),\, h_{\text{textLabelSim}}(C), \\ & h_{\text{entityVDSim}}(C),\, h^A_{\text{hierarchy}}(C), \\ & h^A_{\text{hierarchyProp}}(C),\, h^A_{\text{crissCross}}(C)), \\ & (0.1,\, 0.5,\, 0.3,\, 0.2,\, 0.2,\, 0.4,\, 0.1)) \end{aligned}$$ which is also used as support heuristic.	

Table **B.3.**: Configuration parameters of MapEVO for OAEI 2011.5.

Parameter	Value	Description
i_{max}	$100+$ $\max\{\sharp\mathcal{O}_1,\ \sharp\mathcal{O}_2\}$	number of iterations
$\lvert I\rvert$	32	population size
enable classes	`true`	
enable object properties	`false`	
enable data properties	`false`	
enable individuals	`false`	
κ	0.6	keep-set threshold (cf. Equation (5.19))
σ	0.1	safe-set threshold (cf. Equation (5.20))
topology	`StarTopology`	social network structure (cf. Table 6.5)
comm. strategy	synchronous	cf. Table 6.5
β	0.1	proportional likelihood increment (personal best)
γ	0.05	proportional likelihood increment (local best)
λ	0.3	size change probability factor (cf. Equation (5.24))
w_{inc} start value	2.5	cf. Equation (5.25)
w_{inc} end value	1.5	cf. Equation (5.25)
w_{dec} start value	1.5	cf. Equation (5.25)
w_{dec} end value	1.1	cf. Equation (5.25)
objective function	$$\begin{aligned} F(A) \ =\ & \Gamma_{\text{weightAvg}}((H_{\text{size}}(A),\ H_{\text{corrContrib}}(A),\\ & H_{\text{crissCross}}(A),\ H_{\text{structPreserv}}(A)),\\ & (0.1,\ 0.3,\ 0.1,\ 0.1)) \end{aligned}$$ where for all $C \in A$ $$\begin{aligned} \iota(C) \ =\ & \Gamma_{\text{weightAvg}}((h_{\text{lexIDSim}}(C),\\ & h_{\text{lexLabelSim}}(C),\ h_{\text{textLabelSim}}(C),\\ & h_{\text{entityVDSim}}(C),\ h^A_{\text{hierarchy}}(C),\\ & h^A_{\text{hierarchyProp}}(C),\ h^A_{\text{crissCross}}(C)),\\ & (0.1,\ 0.5,\ 0.3,\ 0.2,\ 0.2,\ 0.4,\ 0.1)) \end{aligned}$$ which is also used as support heuristic.

Table **B.4.**: Configuration parameters of MapPSO for OAEI 2011.5.

Index